SpringerBriefs in Law

SpringerBriefs in Law and Regulation & Innovation

SpringerBriefs present concise summaries of cutting-edge research and practical applications across a wide spectrum of fields. Featuring compact volumes of 50 to 125 pages, the series covers a range of content from professional to academic. Typical topics might include:

- A timely report of state-of-the art analytical techniques
- A bridge between new research results, as published in journal articles, and a contextual literature review
- A snapshot of a hot or emerging topic
- A presentation of core concepts that students must understand in order to make independent contributions

SpringerBriefs in Law showcase emerging theory, empirical research, and practical application in Law from a global author community. SpringerBriefs are characterized by fast, global electronic dissemination, standard publishing contracts, standardized manuscript preparation and formatting guidelines, and expedited production schedules.

SpringerBriefs in Law and Regulation & Innovation present summaries of cutting-edge research and practical applications across the wide spectrum of regulatory law. Regulation has established itself as an important component of the legal system. It aims to achieve the common good in the interaction between and among administration and private individuals and occurs alongside the other central (administrative) legal functions of limiting state power to protect freedom, averting dangers, regulating the direct provision of state services and the general provision of legal rules for social development. SpringerBriefs in Law and Regulation & Innovation are dedicated to the understanding, systematic recording and dogmatic and empirical processing of such regulatory regimes. It is open for basic questions, central elements and new developments in the regulatory law. The focus is on legal regimes to secure results that promote the common good from and through economic competition, such as those found in network economies, data law, and increasingly in the food as well as in the health sector etc. Featuring compact volumes of 50 to 125 pages, the series covers a range of content from mainly professionals but also academics.

Federica Ronchetti • Laura Springer
Kai P. Purnhagen

The Regulatory Landscape in the EU for Dairy Products Derived from Precision Fermentation

An Analysis on the Example of Cheese

Springer

Federica Ronchetti
Faculty of Life Sciences: Food, Nutrition and Health. Chair of Food Law
University of Bayreuth
Kulmbach, Germany

Laura Springer
Faculty of Life Sciences: Food, Nutrition and Health. Chair of Food Law
University of Bayreuth
Kulmbach, Germany

Kai P. Purnhagen
Faculty of Life Sciences: Food, Nutrition and Health. Chair of Food Law
University of Bayreuth
Kulmbach, Bayern, Germany

ISSN 2192-855X ISSN 2192-8568 (electronic)
SpringerBriefs in Law
ISSN 3004-9210 ISSN 3004-9229 (electronic)
SpringerBriefs in Law and Regulation & InnovationISBN
978-3-031-49691-2 ISBN 978-3-031-49692-9 (eBook)
https://doi.org/10.1007/978-3-031-49692-9

This Springer imprint is published by the registered company Springer Nature Switzerland AG
The registered company address is: Gewerbestrasse 11, 6330 Cham, Switzerland

Paper in this product is recyclable

Acknowledgments

The funding for this report was generously provided by the Adalbert-Raps-Foundation, and it was collaboratively developed by the Chair of Food Law within the Faculty of Life Sciences: Food, Nutrition, and Health (Campus Kulmbach), together with the Research Center for German and European Food Law (Forschungsstelle für Deutsches und Europäisches Lebensmittelrecht) at the University of Bayreuth. Additionally, we want to thank David Stöttner for helping us with the finalisation of this project.

Chair of
Food Law
Lehrstuhl für
Lebensmittelrecht

UNIVERSITÄT
BAYREUTH

FLMR
Forschungsstelle für
Deutsches und Europäisches Lebensmittelrecht der
Universität Bayreuth

ADALBERT
RAPS
STIFTUNG

Contents

About the Authors

Federica Ronchetti works as food law expert at the Competence center for the Food Chain of the Austrian Agency for Health and Food Safety (AGES) since September 2023. She contributed to this book while she was a scientific employee at the Chair of Food Law, University of Bayreuth, Campus Kulmbach and the Research Center for Food law of the University of Bayreuth since 2020, where she focused on interdisciplinary studies in the area of food law, in particular on the regulation of food innovations in the EU. She holds a BSc in Nutritional Sciences from the University of Vienna, a Dipl.-Ing. in Safety in the food chain from the University of Natural Resources and Life Sciences, Vienna and an LL.M. in Food law from Luiss Guido Carli, Rome. In 2022, she co-founded the "Bayreuth-Kulmbach Alt Protein Project" of the Good Food Institute.

Laura Springer works as scientific employee at the Chair of Food Law, University of Bayreuth, Campus Kulmbach since 2021 with a focus on Food Law, Food Waste Law and European Law. She holds the German "Erstes Staatsexamen", a law degree which she completed in Bayreuth in 2021.

Kai P. Purnhagen is Full Professor and Director of the Research Center for German and European Food Law at the University of Bayreuth. He works in the areas of European Union Law, International Economic Law and the interdisciplinary and comparative analysis of law, with emphasis on the food market. He studied Law, Political Science, Economics, Philosophy and Sociology at the Justus-Liebig-University Gießen and at the Fernuniversity Hagen, he continued as LL.M. - Fellow at the University of Wisconsin-Madison, before he was admitted to the PhD programme of the European University Institute in Florence, he undertook a Post-doc at the University of Amsterdam and at the Ludwig-Maximilians-University Munich, before he was appointed Assistant and Associate Professor in Law at Wageningen University. He held visiting or short-term appointments at the University of Lucerne, the London School of Economics and Political Science, the University of Wisconsin-Madison, and the Justus-Liebig-University Giessen.

Abbreviations

ANS Panel	EFSA's Panel on Food Additives and Nutrient sources added to Food
BIOHAZ Panel	EFSA's Panel on Biological Hazards
CJEU	Court of Justice of the European Union
CEF Panel	EFSA Panel of Food Contact Material, Enzymes, Flavourings and Processing Aids Panel
CMO	Common Market Organisation
DG SANTE	Commission's Directorate-General for Health and Food Safety
DNA	Desoxyribonucleic acid
e.g.	exempli gratia
EC	European Commission
ECJ	European Court of Justice
EFSA	European Food Safety Authority
EU	European Union
EVU	European Vegan Union
FAR	Food Additives Regulation
FBO	Food business operator
FD&C Act	Federal Food, Drug, and Cosmetic Act
FDA	U.S. Food and Drug Administration
FEEDAP Panel	EFSA's Panel on Additives and Products or Substances used in Animal Feed
FIR	Food Information Regulation
GFL	General Food Law
GM	Genetically modified
GMF	Genetically modified food
GMFR	Genetically Modified Food and Feed Regulation
GMM	Genetically modified microorganism
GMO Panel	EFSA's Panel on Genetically Modified Organisms
GMO	Genetically modified organism
GRAS	Generally recognized as safe
HNCR	Regulation on Nutrition and Health Claims

i.e.	id est
min.	minutes
NFR	Novel Food Regulation
PDO	Protected designation of origin
PGI	Protected geographical indication
QPS	Qualified Presumption of Safety
rDNA	Recombinant DNA
S.	Sentence
SDG	Sustainable Development Goal
SFA	Singaporean Food Authority
Sub.	Subparagraph
TEU	Treaty on European Union
TFEU	Treaty on the Functioning of the European Union
UN	United Nations
USA	United States of America

Chapter 1
Executive Summary

This report assesses the EU legal framework applicable to dairy products obtained through precision fermentation. It maps the authorisation and labelling requirements for these products to be placed on the EU market. It compares these provisions to those governing the food markets in the USA and Singapore, which are considered to be more innovation-friendly. At the time of writing, the precision-fermented dairy sector is still in the developmental phase, with only a few precision-fermented dairy alternatives available on the market, none of which are present in the EU. Regulatory uncertainty has been cited as significant obstacle for food business operators seeking to introduce alternative protein products in the EU. Our research confirms these reports, as we found that the current EU regulatory framework presents several legal uncertainties that are challenging for food business operators to overcome.

Broadly speaking, there are two authorisation frameworks applicable to dairy products obtained through precision fermentation in the EU. The choice between these frameworks depends on the presence of genetically modified organisms (GMOs) or residues thereof in the final product. If such GMOs or residues are present, the pre-market authorisation procedure is governed by the Genetically Modified Food and Feed Regulation (GMFR). If the final product does not contain any GMOs or residues, it falls under the scope of the Novel Food Regulation (NFR), and the pre-market authorisation procedure is governed accordingly. Both of these authorisation pathways entail high regulatory requirements, which can pose challenges for the development of the precision fermentation sector, particularly for start-ups and small to medium-sized enterprises. Despite the difficulty in accurately classifying products, it is crucial for food business operators to comprehend the applicable legal framework early in the product development process, given that the authorisation paths differ in their standards. Consequently, many companies operating in this sector may prefer to focus on other, more accessible markets.

F. Ronchetti et al., *The Regulatory Landscape in the EU for Dairy Products Derived from Precision Fermentation*, SpringerBriefs in Law,
https://doi.org/10.1007/978-3-031-49692-9_1

As part of the pre-market authorisation requirements, companies must also address labelling issues. Food labelling shall provide a basis for consumers to make informed choices in relation to the foods they consume. The primary objective of EU food labelling law is to prevent consumer deception,[1] making it crucial for dairy products and their animal-free alternatives, to adhere to a stringent naming law as stipulated in the Common Market Organisation and specific accompanying legislation.

Moreover, it is imperative to consider regulations regarding the use of health and nutrition claims, as well as labels such as “vegan” or “organic”, along with clean labels and other private certification schemes. Furthermore, the labelling of novel foods and genetically modified foods may necessitate mandatory information specified in the authorisation process. These labelling requirements demand close attention to ensure compliance and to provide accurate and transparent information to consumers.

Our research serves as a comprehensive guideline for food business operators involved in the production of precision-fermented dairy alternatives, aiding them in navigating the intricate regulatory landscape in the EU. However, it is important to note that our research specifically concentrates on the legal implications directly associated with the use of precision fermentation in food production.

The marketing of the final product may entail individualized counselling on a case-by-case basis, particularly concerning the presence of additional components added to the food. As such, it is essential for companies to seek tailored guidance to ensure compliance with all relevant regulations when bringing their products to market.

[1] Article 1(1) Regulation (EU) No 1169/2011 (2011).

Chapter 2
Background

On the 15th of November 2022 the world population surpassed the 8-billion mark. Predictions expect the population to reach the 10.4 billion mark in the 2080s.[1] A growing population correlates with an increasing need for resources.

The United Nations (UN) is well aware of this issue and has addressed it in the 2030 Agenda for Sustainable Development.[2] This agenda includes 17 Goals and 169 Targets attempting to balance the three dimensions of sustainable development: the economic, the social and the environmental level.[3] Sustainable Development Goal (SDG) 2 concerns ending hunger, achieving food security and improving nutrition and promoting sustainable agriculture. This goal is divided into five targets. Target 2.4 refers to a sustainable food production and implements resilient agricultural practices that increase productivity and production, that help maintain ecosystems and strengthen capacities for adaptation to climate change.[4] The question is: How can our food production be sustainable and still be able to feed the world population and end hunger?

Several governments have begun incorporating the concept of sustainability into their agendas. In the EU, the European Commission has adopted the "Green Deal." Part of the "Green Deal" is the "farm-to-fork strategy,"[5] stipulating that the establishment of a sustainable food supply chain is one of the main goals of the EU. In its report on the transition to sustainable food, the European Consumer Organisation recommends promoting and supporting consumers in adopting more plant-based diets. The report also suggests making attractive and affordable alternative protein sources available to the public.[6]

The consumption of dairy—especially cheese products—is growing annually. In 2022 around 21.3 million tons of cheese were consumed worldwide; with an increasing trend.[7] If consumption of dairy products continues to increase,

---

[1] United Nations - Department of Economic and Social Affairs (2022).

[2] United Nations - General Assembly (2015).

[3] United Nations - General Assembly (2015).

[4] United Nations - General Assembly (2015).

[5] European Commission (2019, 2020).

[6] The European Consumer Organisation (2020).

[7] United States Department of Agriculture (2023).

F. Ronchetti et al., *The Regulatory Landscape in the EU for Dairy Products Derived from Precision Fermentation*, SpringerBriefs in Law,
https://doi.org/10.1007/978-3-031-49692-9_2

alternative dairy production processes may be necessary to offset the negative environmental impacts caused by traditional dairy production.

In the past decades, new technologies disrupted the food market: At the beginning of the new millennium, lab grown meat was considered science fiction. However, in 2013, the first cultured beef burger was unveiled by Mosa Meat.[8] The production of one burger is yet too expensive to meet market demands and the development of cultured meat still has a long way to go before it is ready for the food market. Other types of technological innovations in the food sector are already placed on the market and promise change of our diet in the future. In addition to changes in consumer habits such as vegan diets or the consumption of insects, some new technological innovations are already suitable for mass production. The production of foods obtained via precision fermentation is a promising technology which can be available for successful scaling up. In the last years, an increasing amount of companies started to work with precision fermentation, targeting different foodstuffs (e.g. cheese, eggs or even breast-milk). Precision fermentation is considered to having the potential to be a "game changer" when it comes to producing food in a sustainable way. Nevertheless, since the employment of precision fermentation is relatively new in food production, especially on the EU market, many regulatory questions such as those concerning authorisation, classification, and labelling, remain unanswered.

2.1 Problem Description and Aim of the Project

EU law imposes comparably high standards when it comes to placing innovations on the food market. Next to the establishment of the internal market, health and food safety are major policy objectives of EU food law, as provided by Article 1 of the General Food Law (GFL).[9] In practice, health and safety are the major goals of EU food law. This emphasis on health and safety results in cumbersome legal procedures for food business operators (FBOs) wanting to place new food innovations on the EU market. While elaborate procedures and lengthy review processes may serve health and safety of food innovations, they also reflect in the company's costs, so that companies are hesitant when it comes to investing in food innovations.[10]

Policies and regulations have a twofold effect on food innovations, as they can both induce and promote their development by offering them an insurance of return of investments, but also hamper their advancement by establishing stringent rules and procedures.[11] Considering the critical role that a clear and predictable

[8] MosaMeat (2023).

[9] Regulation (EC) No 178/2002 (2002).

[10] Holle (2018); Molitorisová et al. (2021).

[11] Eriksson et al. (2020a, b, c); Finck (2018).

regulatory framework can play in facilitating innovation, businesses may feel uncertain how when to profit from this insurance function by granting market access.

This report provides a comprehensive overview of the (EU) laws applicable to dairy alternatives obtained via precision fermentation. We divided the report into three parts:

- I Authorisation. This section focusses on the legal classification of products derived from precision fermentation in the EU. It identifies the applicable legal framework and compares the novel food and the GMO pre-market authorisation procedures.
- II Labelling. This section provides an overview of the labelling requirements of precision fermented dairy alternatives, including questions of denomination, the provision of mandatory information and the use of voluntary claims and certification schemes.
- III Broad comparative analysis. This section compares authorisation frameworks of precision fermented foods in other countries (i.e. Singapore and the USA) and with other products.
  This report will not serve as a guideline for the final marketing of the respective food products. Instead, it will focus on examining the regulatory challenges associated with precision fermentation in cheese and dairy production, specifically in relation to the synthesis of casein. The scope of this report does not encompass aspects concerning further production steps and downstream processing, including the addition of food improvement agents (food enzymes, additives, or flavourings) and other ingredients.

2.2 Methodology

We examined relevant EU regulations and directives, primary EU law, case law of the Court of Justice of the European Union, literature, as well as relevant (soft) law guidance at the national, European, and international level. By carrying out interviews with relevant stakeholders we combined doctrinal legal analysis with empirical research.

2.2.1 Doctrinal Legal Research

We have examined several EU law documents that provide us with information on the classification, authorisation and labelling of products derived from precision fermentation.

In conducting the research, we established connections between various relevant parts of EU law and case law. This involved employing multiple interpretative methods, including systematic and teleological interpretation. Throughout the process of the research, we relied on doctrinal legal research. The doctrinal legal analysis has been serving as the foundational framework for formulating the empirical legal research.

2.2.2 Empirical Legal Research

Our empirical legal research part involved qualitative interviews with relevant stakeholders, including companies and consultants in this field, who shared insights on the EU regulatory environment's operations. These interviews focused on their experiences with precision fermentation and the EU food innovation regulatory environment.

We conducted seven semi-structured expert interviews via Zoom, with each interview lasting up to 45 min. During these sessions, ample opportunities were provided for the interviewees to share their thoughts and express concerns or doubts they had regarding the regulatory environment. Fortunately, the Adalbert-Raps-Stiftung has generously supplied us with all the essential contacts. Due to confidentiality agreements, our interviewees and their companies will be kept anonymous and will not be referenced. The majority of our interviewees work in start-ups, mostly in the developing phase. Some regulatory experts (such as consultants for companies in the food innovation sector) were also interviewed. The interviews provided first-hand information about the challenges companies in the precision fermentation sector are facing when it comes to launching the products on the market.

2.2.3 Limitations

Our empirical legal research encountered several limitations that had an impact on the study. Most companies working with precision fermentation are still in the developing phase. While they possessed substantial knowledge about processing their products, they lacked a comprehensive understanding of the regulatory framework they would face when introducing their products on the EU market. Due to the stringent regulatory environment in the EU, companies that were on the verge of launching their products tended to be based in the USA or Singapore, and they were not (yet) actively operating within the EU market. As a result, obtaining direct insights and data from companies currently active in the EU market proved challenging.

Considering that these companies are relatively new and there are currently no precision-fermented dairy alternative products placed on the EU market, consumption and large-scale production of such products have not yet occurred. As a result, the need for further research becomes apparent, particularly as it is expected that these products will enter the market in the upcoming years. At this stage, some of the challenges that companies may face cannot be predicted with certainty.

Therefore, our research primarily focuses on comprehending and addressing the pre-market steps involved in this innovative sector. Additional research will be crucial in the coming years, especially when precision fermentation products are introduced to the market. Many challenges that companies may encounter cannot be predicted at this stage. Given these uncertainties, our research primarily centres on comprehending and addressing the pre-market steps involved in this innovative sector. Another issue that arose during the research pertained to the reluctance of most start-ups to cooperate and share essential insights that would have significantly contributed to shaping this report. The majority of these start-ups were not yet at a stage where they could launch their products in the EU market, and some have not even begun the authorisation process with EFSA or with the competent national authorities. As a result, certain interviewees expressed that they were not prepared or able to provide specific insights, while others had not yet considered potential regulatory challenges. Additionally, a few interviewees struggled to comprehend the legal context and were unaware of the legal hurdles they might encounter. This led to that lacked thorough consideration of legal questions.

On a positive note, certain consultancies in the novel food sector displayed a commendable willingness to share valuable information about the challenges their clients encounter concerning the authorisation and labelling of their products. This cooperative attitude proved to be incredibly helpful for our research. However, the prevailing status quo underscores the significance of this project for numerous start-ups in the precision fermentation sector. The majority of these start-ups remain uncertain about the potential regulatory challenges they might face in the future. As a result, the insights gained from this research hold immense value in providing guidance and clarity to these emerging companies.

2.2.4 Interview Results: Producer Expectations

Initially we familiarised ourselves with the interviewees and their respective companies. This allowed us to gain valuable insights into how they became involved with precision fermentation.

Given the broad definition of precision fermentation, we asked each interviewee to provide their own interpretation of the term. Subsequently we delved into classification issues, seeking to understand which authorisation process the interviewees had applied or planned to apply in the EU. Interestingly, most interview partners were uncertain whether their product is covered by the novel food or the GMO regulatory framework.

Furthermore, we inquired about the intended labelling of their products. Given the diversity within the precision fermentation sector and the early developing status of many companies, market entry emerged as their primary concern, with the regulatory focus mainly on the authorisation process. While most interviewees did not give much thought to labelling, they were open to share their ideas. Particularly when it came to product names, opinions among the interviewees varied widely:

Some were confident that they could use conventional names (e.g.: "cheese") for their precision-fermented alternatives, while others acknowledged the limitations and the need for unique product denominations. The same applies to further certifications of their products, e.g. as vegan or organic.

During the closing remarks, the interviewees engaged in discussions about future challenges and expectations for the precision fermentation sector both within the EU and on a global scale. Each interviewee emphasised the complexity of the EU legal framework and the bureaucratic burden they face within the EU administration. They expressed that navigating through the regulations and processes can be a challenging endeavour.

Interestingly, every interviewee highlighted the USA and Singapore as role models for innovative food and markets, perceiving them to offer easier access and more supportive environments for their businesses. They considered these countries to be leading the way in fostering innovation in the food industry.

Moreover, the interviewees perceived EU consumers to be more cautious on average when it comes to accepting and embracing innovative food products or producing methods. This cautiousness was seen as a potential challenge in introducing precision-fermented products to the EU market, as gaining consumer acceptance may require additional effort.

Regarding products derived from precision fermentation, the interview partners were optimistic. They expressed confidence that these products would soon become available in the EU market due their safety of consumption. This optimism stems from the belief that precision fermentation can offer sustainable and safe alternatives to traditional food products.

The results of the interviews are included in this report.

2.3 Legal Framework

This section presents a list of the most frequently used legislative acts and a brief description of their content.

European Primary Law: Secondary legislative acts of the EU are based on the competence norms enshrined in the Treaty on the Functioning of the European Union (TFEU) and the Treaty on European Union (TEU). These treaties are part of the EU primary law. They are binding agreements that set out, inter alia, the objectives of the EU. All EU regulations and directives are termed as secondary law and they must be in conformity with primary law.[12] Primary law builds the foundation of the laws that will be part of this book, and one of its most relevant objectives for the scope of this book is the achievement of sustainability. Article 3(2)(Sub. 1)(2) TEU stipulates that the sustainable development is the base for an economic growth in Europe. In addition, Article 11 TFEU highlights that environmental protection

[12] European Commission (2023).

requirements must be integrated into the definition and implementation of the Union's policies and activities. In essence, Article 37 of the Charter reiterates the content of Article 11 TEU and does not add any new significance to it.[13]

General Food Law (GFL): Regulation (EC) No 178/2002 of the European Parliament and of the Council of 28 January 2002 laying down the general principles and requirements of food law, establishing the European Food Safety Authority and laying down procedures in matters of food safety, OJ L 031, 01.02.2002, pp. 1–24.

The GFL constitutes the core of the European food legislation. It contains the general principles and definitions of European food law and applies directly in all Member States.[14] Next to the definition of food in Article 2 GFL, the GFL provides for the core requirement for food safety. According to Article 14(1) GFL, food that is unsafe shall not be placed on the market. Paragraph 2 clarifies that food is considered unsafe if it is injurious to health or unfit for human consumption. This ban applies to any placing on the market of the food, at any stage in the food supply chain.[15] The control over the compliance with all legal requirements is the responsibility of the food business operator according to Article 17(1) GFL. His obligations are not limited to the provisions of the GFL and can arise from all food law rules.

Food Information Regulation (FIR): Regulation (EU) No 1169/2011 of the European Parliament and of the Council of 25 October 2011 on the provision of food information to consumers, amending Regulations (EC) No 1924/2006 and (EC) No 1925/2006 of the European Parliament and of the Council, and repealing Commission Directive 87/250/EEC, Council Directive 90/496/EEC, Commission Directive 1999/10/EC, Directive 2000/13/EC of the European Parliament and of the Council, Commission Directives 2002/67/EC and 2008/5/EC and Commission Regulation (EC) No 608/2004; OJ L 304, 22.11.2011, pp. 18–63.

In addition to the GFL, the FIR also lays down essential principles for the handling of foodstuffs. Its Article 1 provides the basis for the assurance of a high level of consumer protection in relation to food information, which must be fulfilled by comprehensive labelling obligations. Although the FIR gives EU labelling law overriding priority over national law, a number if its provisions require interpretation. European Commission and national authorities' guidance documents aid with this interpretation. Article 9 FIR serves as the basic norm for mandatory labelling requirements (e.g. name of the food list of ingredients or the nutrition declaration) whereas Article 36 FIR includes rules for any voluntary food information.

Novel Food Regulation (NFR): Regulation (EU) 2015/2283 of the European Parliament and of the Council of 25 November 2015 on novel foods, amending Regulation (EU) No 1169/2011 of the European Parliament and of the Council and repealing Regulation (EC) No 258/97 of the European Parliament and of the Council and Commission Regulation (EC) No 1852/2001; OJ L 327, 11.12.2015, pp. 1–22.

---

[13] Epiney (2019), p. 49 marginal no. 3.

[14] Rützler (2023).

[15] Wirbel (2021).

The NFR regulates novel foods. Article 4(1) NFR stipulates that food business operators shall verify whether the food which they intend to place on the market within the Union falls within the scope of the NFR. It is important to know whether the product intended to be placed on the market is a novel food, because novel foods require authorisation (see Articles 10 ff. NFR).

Genetically Modified Food and Feed Regulation (GMFR): Regulation (EC) No 1829/2003 of the European Parliament and of the Council of 22 September 2003 on genetically modified food and feed; OJ L 268, 18.10.2003, pp. 1–23.

Genetically Modified Food (GMF) is, according to Article 2(1) No. 6 GMFR, any food that contains, consists or is produced from genetically modified organisms (GMO). GMF is subject to a pre-market authorisation process and other post-market regulations such as specific labelling requirements.

Common Market Organisation (CMO): Regulation (EU) No 1308/2013 of the European Parliament and of the Council of 17 December 2013 establishing a common organisation of the markets in agricultural products and repealing Council Regulations (EEC) No 922/72, (EEC) No 234/79, (EC) No 1037/2001 and (EC) No 1234/2007.

The scope of the Regulation is to establish a common organisation of the markets for agricultural products, which includes milk and milk products (see Article 1 CMO). Especially Annex VII, Part III contains important information about the labelling and naming of milk and milk products.

2.4 Definitions

This section contains the definitions of the key terms used in this report. Since food law is in many areas subject to maximum harmonisation,[16] most definitions are directly provided for in EU legislation.

(Traditional) fermentation is a technology used in food production whereby organic substrates are chemically broken down by microbial activity to obtain desirable changes in the product.[17] Food fermentation is a common process and has a long history of use. Food had been fermented long time before any systematic knowledge on biochemistry and microbiology was available because it enhances preservation and induces a change in the organoleptic properties of the foodstuffs.[18] The progress of science now enables direct and precise adjustments in the fermentation process to fulfil specific purposes.[19]

There is currently no officially legal or commonly accepted scientific definition of **precision fermentation**. However, during our interviews, stakeholders often

[16] Sachs (2016), p. 410.

[17] Hui et al. (2016).

[18] Campbell-Platt (1994).

[19] Ojeda-Linares et al. (2021).

referred to it as a process where single-celled organisms are harnessed as "living factories" and cultivated under specific conditions to produce a target by-product.

In some cases, microorganisms that would normally not produce the target compound can be subjected to bioengineering. For example, milk proteins are not naturally produced by single-celled organisms. To overcome this hurdle, the genes coding for certain milk proteins (i.e., caseins) are integrated into the genome of a suitable microorganism. The genetically modified microorganism is cultivated under controlled conditions, so that the appropriate gene expression is induced, and casein production takes place. This bioidentical casein is then harvested and purified, and finally used for further processing (i.e., cheese production).

The definition for **food** is stated in Article 2(Sub. 1) GFL:

> For the purposes of this Regulation, 'food' (or 'foodstuff') means any substance or product, whether processed, partially processed or unprocessed, intended to be, or reasonably expected to be ingested by humans.

Sub. 2 names substances which explicitly fall under the food definition, e.g. beverages. Instead, Sub. 3 explicitly excludes categories from the definition. This negative definition includes, for example, animal feed (Sub. 3(a)). The definition of food is a broad one to allow a wide scope of application of the GFL.[20] Whenever a good falls under the definition of food, it is mandatory to apply the EU food law requirements for a placing of this good on the market in the EU.[21]

Article 3(2)(a) of the Regulation on **Food Additives** (FAR)[22] defines food additives as "any substance not normally consumed as a food in itself and not normally used as a characteristic ingredient of food, whether or not it has nutritive value, the intentional addition of which to food for a technological purpose in the manufacture, processing, preparation, treatment, packaging, transport or storage of such food results, or may be reasonably expected to result, in it or its by-products becoming directly or indirectly a component of such foods."

Article 3 FAR provides a list of substances which are not considered to be food additives. This list includes for example milk proteins (No. viii) and caseinates and casein.

Processing aids are defined in Article 3(2)(b) FAR as "any substance which (i) is not consumed as a food by itself; (ii) is intentionally used in the processing of raw materials, foods or their ingredients, to fulfil a certain technological purpose during treatment or processing; and (iii) may result in the unintentional but technically unavoidable presence in the final product of residues of the substance or its derivatives provided they do not present any health risk and do not have any technological effect on the final product."

Contrasting additives, no or only unintentional and technically unavoidable residues of processing aids remain in the final product, whereas additives (can) become

[20] Meisterernst (2019), § 4 marginal no. 2.

[21] Molitorisová et al. (2021), p. 15.

[22] Regulation (EC) No 1333/2008 (2008).

a component of the foodstuff. For this reason, a processing aid does not need to be included in the list of ingredients of the product.[23]

Food enzymes are defined in Article 3(2)(a) Regulation (EC) No 1332/2008[24] on food enzymes as "a product obtained from plants, animals or micro-organisms or products thereof including a product obtained by a fermentation process using micro-organisms: (i) containing one or more enzymes capable of catalysing a specific biochemical reaction; and (ii) added to food for a technological purpose at any stage of the manufacturing, processing, preparation, treatment, packaging, transport or storage of foods." According to Article 4 of the Regulation only food enzymes included in the Community list may be placed on the market.

Any food that was not used for human consumption to a significant degree within the Union before 15 May 1997, irrespective of the dates of accession of Member States to the Union and can be categorised in one of the four novel food categories per Article 3(2)(a) Novel Food Regulation (NFR) is referred to as **novel food**. These foods are prohibited for marketing in the EU, unless explicitly permitted, which means that Novel foods can only be placed on the market after their authorisation and inclusion in an EU list for novel foods, see Article 6(2) NFR.

Organisms, with the exception of human beings, in which the genetic material has been altered in a way that does not occur naturally by mating and/or natural recombination are called **genetically modified organisms**, see Article 2 No. 5 Regulation (EC) No 1829/2003 (GMFR) and Article 2 No. 2 of Directive 2001/18/EC. Annex I B to this Directive contains a list of excluded organisms. The authorisation process for GMO differs from the one for novel foods. However, Member States have an "opt-out" possibility for cultivation. This means that a Member State can withdraw the permission to cultivate GMO-products on its territory, Article 23 of Directive 2001/18/EC.

Casein is a glycoprotein and the major protein found in milk, accounting for 80% of its protein content. It includes four protein fractions (α1-, β2-, b- and k-casein), which are organised in so-called micelles.[25] In 2015 the EU adopted Directive (EU) 2015/2203 on the approximation of the laws of the Member States related to caseins and caseinates intended for human consumption, setting out common standards (essential factors of composition, maximum levels for contaminants and impurities, allowed processing aids, bacterial cultures and authorised ingredients, and organoleptic characteristics[26]).

Milk is legally defined as "the normal mammary secretion obtained from one or more milkings without either addition thereto or extraction therefrom" in Annex VII Part III, 1 CMO. It is thereby assumed that the milk has bovine origin. Should it originate from another animal species, its origin must be stated (Annex VII Part III, (4)). The CMO also stipulates that the term "milk" may be used exclusively for milk

---

[23] Niedersächsisches Landesamt für Verbraucherschutz.

[24] Regulation (EC) No 1332/2008 (2008).

[25] McMahon and Oommen (2013); Ptiček Siročić (2017).

[26] Annex I. Directive (EU) 2015/2203 (2015).

treated without altering its composition or for milk with a standardized fat content (Annex VII Part III, (1)(a)).

The definition of **dairy products** in EU law is very restrictive. According to the CMO (Annex VII Part III, 2), the category "**milk products**" includes "products derived exclusively from milk, on the understanding that substances necessary for their manufacture may be added provided that those substances are not used for the purpose of replacing, in whole or in part, any milk constituent". Some specific terms are reserved exclusively for "milk products".

References

Campbell-Platt G (1994) Fermented foods — a world perspective. Food Res Int 27:253–257. https://doi.org/10.1016/0963-9969(94)90093-0

Epiney A (2019) Umweltrecht der Europäischen Union, 4th edn. Beck-Online Bücher. Nomos; facultas, Baden-Baden, Wien

Eriksson D, Custers R, Edvardsson Björnberg K, Hansson SO, Purnhagen K, Qaim M, Romeis J, Schiemann J, Schleissing S, Tosun J, Visser RGF (2020a) Options to reform the European Union legislation on GMOs: post-authorization and beyond. Trends Biotechnol 38:465–467. https://doi.org/10.1016/j.tibtech.2019.12.015

Eriksson D, Custers R, Edvardsson Björnberg K, Hansson SO, Purnhagen K, Qaim M, Romeis J, Schiemann J, Schleissing S, Tosun J, Visser RGF (2020b) Options to reform the European Union legislation on GMOs: risk governance. Trends Biotechnol 38:349–351. https://doi.org/10.1016/j.tibtech.2019.12.016

Eriksson D, Custers R, Edvardsson Björnberg K, Hansson SO, Purnhagen K, Qaim M, Romeis J, Schiemann J, Schleissing S, Tosun J, Visser RGF (2020c) Options to reform the European Union legislation on GMOs: scope and definitions. Trends Biotechnol 38:231–234. https://doi.org/10.1016/j.tibtech.2019.12.002

European Commission (2019) The European Green Deal. https://eur-lex.europa.eu/legal-content/EN/TXT/?uri=celex%3A52019DC0640. Accessed 1 January 2024

European Commission (2020) A Farm to Fork Strategy for a fair, healthy and environmentally-friendly food system. https://eur-lex.europa.eu/legal-content/EN/TXT/?uri=CELEX%3A52020DC0381. Accessed 1 January 2024

European Commission (2023) Types of EU law. https://commission.europa.eu/law/law-making-process/types-eu-law_en. Accessed 9 June 2023

Finck M (2018) Blockchain regulation and governance in Europe. Cambridge University Press

Holle M (2018) Pre-market approval and its impact on food innovation: the novel foods example. In: Bremmers H, Purnhagen K (eds) Regulating and managing food safety in the EU. Springer International Publishing, Cham, pp 291–330

Hui YH, Evranuz EÖ, Bingöl G, Erten H, Jaramillo-Flores ME (eds) (2016) Handbook of vegetable preservation and processing. CRC Press Taylor & Francis Group, Boca Raton, London, New York

McMahon DJ, Oommen BS (2013) Casein micelle structure, functions, and interactions. In: McSweeney PLH, Fox PF (eds) Advanced dairy chemistry. Springer US, Boston, pp 185–209

Meisterernst A (2019) Lebensmittelrecht. Studium und Praxis. C.H. Beck, München

Molitorisová A, Monaco A, Purnhagen KP (2021) An analysis of the regulatory framework applicable to products obtained from mushroom and mycelium. SSRN J. https://doi.org/10.2139/ssrn.3955899

MosaMeat (2023) The mission: to fundamentally reshape the global food system. https://mosameat.com/the-mission. Accessed 16 June 2023

Niedersächsisches Landesamt für Verbraucherschutz Etikett: Alles (un)klar?: Clean Labelling: Information Oder Täuschung? https://www.laves.niedersachsen.de/startseite/lebensmittel/kennzeichnung/etikett-alles-unklar-110019.html. Accessed 9 June 2023

Ojeda-Linares C, Álvarez-Ríos GD, Figueredo-Urbina CJ, Islas LA, Lappe-Oliveras P, Nabhan GP, Torres-García I, Vallejo M, Casas A (2021) Traditional fermented beverages of Mexico: a biocultural unseen foodscape. Foods 10. https://doi.org/10.3390/foods10102390

Ptiček Siročić A (2017) Characterization of casein fractions – comparison of commercial casein and casein extracted from cow's milk. Chem Biochem Eng Q 30:501–509. https://doi.org/10.15255/CABEQ.2015.2311

Rützler (2023) II. Grundlagen des Lebensmittelrechts. In: Streinz R, Kraus M (eds) Lebensmittelrechts-Handbuch: 44. Auflage

Sachs G (2016) Introduction to European food law and regulation. In: Steier G, Patel KK (eds) International food law and policy. Springer International Publishing, Cham, pp 409–450

The European Consumer Organisation (2020) One bite at a time: consumers and the transition to sustainable food: analysis of a survey of European consumers on attitudes towards sustainable food. https://www.beuc.eu/sites/default/files/publications/beuc-x-2020-042_consumers_and_the_transition_to_sustainable_food.pdf. Accessed 7 Aug 2023

United Nations - Department of Economic and Social Affairs (2022) World population to reach 8 billion on 15 November 2022. https://www.un.org/en/desa/world-population-reach-8-billion-15-november-2022. Accessed 9 June 2023

United Nations - General Assembly (2015) Transforming our world: the 2030 Agenda for Sustainable Development: A/RES/70/1. https://www.un.org/en/development/desa/population/migration/generalassembly/docs/globalcompact/A_RES_70_1_E.pdf. Accessed 9 June 2023

United States Department of Agriculture (2023) Diary: world markets and trade. https://apps.fas.usda.gov/psdonline/circulars/dairy.pdf. Accessed 11 Aug 2023

Wirbel C (2021) Strategien zur Bekämpfung von Lebensmittelverschwendung im Lichte des wohlgeordneten Rechts. Dissertation

Chapter 3
Precision Fermentation and GMMs

To determine the appropriate legal framework for dairy alternatives produced using precision-fermented casein, it is crucial to have a clear understanding of the employed technology and the production steps involved.

Currently, there is no universally accepted or authoritative definition of precision fermentation. While it has been increasingly referenced in scientific and journalistic articles, sometimes referred to as "targeted fermentation,"[1] its comprehensive meaning remains unclarified. Existing definitions found in the literature often lack precision and fail to encompass all relevant aspects of the process. For this report, we have chosen to adopt a working definition derived from an explanation of how precision fermentation operates. This allows us to have a functional understanding of the process while acknowledging the lack of an authoritative definition.

The precision fermentation process relies on harnessing the potential of microflora and its metabolism to generate specific compounds. By meticulously controlling and adjusting the environmental conditions to which a microorganism is exposed (including the temperature and the available substrate), specific by-products can be induced in their metabolic production. Microorganisms are thus transformed into "living factories" capable of producing substantial quantities of a target compound. In certain instances, the target substance is naturally synthetised by the employed microflora, while in most cases, bacteria or yeasts must undergo genetic modification to enable production of non-proprietary substances. Precision fermentation is not a recent innovation. Its utilisation dates back to the 1970s when it was employed to produce human insulin using *Escherichia coli.*[2] Furthermore, the food sector has employed this technology since the 1980ies to produce the enzyme chymosin (rennet).[3] Precision fermented chymosin holds the distinction of being the

[1] Hohlreiter and Naumer (2021); Kateman (2021); ProVeg Incubator (2021); Pua et al. (2022).

[2] Cohen et al. (1973); Mendly-Zambo et al. (2021).

[3] Garg and Johri (1994).

F. Ronchetti et al., *The Regulatory Landscape in the EU for Dairy Products Derived from Precision Fermentation*, SpringerBriefs in Law,
https://doi.org/10.1007/978-3-031-49692-9_3

first genetically engineered product to receive approval for human consumption by the U.S. Food and Drug Administration (FDA).[4] In the EU, Chymosin produced using certain genetically modified *Aspergillus niger* var. awamori strains (DSM 29544, DSM 29545 and DSM 29546) is already included in the "Register of Food Enzymes to be considered for Inclusion in the Union List" of the European Commission's Directorate General for Health and Food Safety.[5] Furthermore, EFSA has conducted a positive safety assessment of this product.[6]

The production of dairy product alternatives through precision fermentation involves the use of genetically engineered microorganisms to produce casein, a protein specific to milk. Start-ups involved with the production of these foodstuffs describe the process as shown in Fig. 3.1[7]

The genetic information required for strain engineering is available in databases, enabling a production process that does not involve animals or animal cells at any stage.[8] According to producers, in addition to ethical benefits, precision fermented

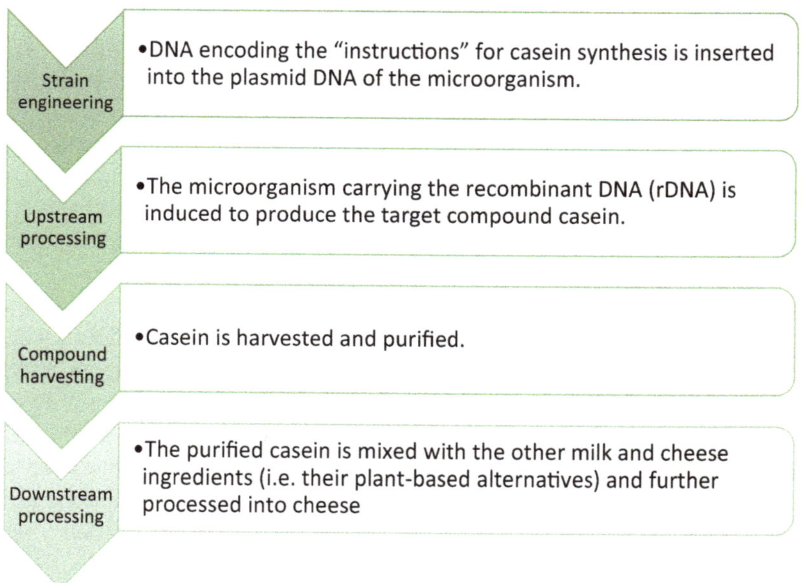

Fig. 3.1 Production of dairy product alternatives through precision fermentation

---

[4] Gladwell (1990).

[5] European Commission (2020).

[6] Silano et al. (2022).

[7] Compton (2016); Mendly-Zambo et al. (2021); Pandya (2014); Tuomisto et al. (2017).

[8] TurtleTree (2023).

dairy also offers several sustainability, safety, and nutrition-related advantages: it can have a longer shelf life, lower contamination potential, and be free from hormones, antibiotics, and lactose.[9]

References

Cohen SN, Chang AC, Boyer HW, Helling RB (1973) Construction of biologically functional bacterial plasmids in vitro. Proc Natl Acad Sci U S A 70:3240–3244. https://doi.org/10.1073/pnas.70.11.3240

Compton NB (2016) This startup is making dairy milk without cows. https://www.vice.com/en/article/vvq33j/this-startup-is-making-dairy-milk-without-cows. Accessed 7 Aug 2023

European Commission (2020) Register of food enzymes to be considered for inclusion in the union list. https://food.ec.europa.eu/system/files/2020-06/fs_food-improvement-agents_enzymes_register.pdf. Accessed 16 June 2023

Garg SK, Johri BN (1994) Rennet: current trends and future research. Food Rev Int 10:313–355. https://doi.org/10.1080/87559129409541005

Gladwell M (1990) FDA approves bioengineered Cheese Enzyme. https://www.washingtonpost.com/archive/politics/1990/03/24/fda-approves-bioengineered-cheese-enzyme/c4292eeb-1c74-45d2-94c3-b0eb09e4866c/. Accessed 16 June 2023

Hohlreiter S, Naumer C (2021) Targeted fermentation for industry. https://easyengineering.eu/targeted-fermentation-for-industry/. Accessed 16 June 2023

Kateman B (2021) Fermentation: the new game-changer for alternative proteins? https://www.forbes.com/sites/briankateman/2021/06/07/fermentation-the-new-game-changer-for-alternative-proteins/?sh=89470233affd. Accessed 16 June 2023

Mendly-Zambo Z, Powell LJ, Newman LL (2021) Dairy 3.0: cellular agriculture and the future of milk. Food Cult Soc 24:675–693. https://doi.org/10.1080/15528014.2021.1888411

Pandya R (2014) Milk without the moo. New Sci 222:28–29. https://doi.org/10.1016/S0262-4079(14)61260-4

Perfect Day (2023) We're rethinking everything. Starting with food.: A kinder, greener protein designed for a sustainable future. http://www.perfectdayfoods.com/. Accessed 7 Aug 2023

ProVeg Incubator (2021) Fermentation: a quick guide for alt-protein fans. https://provegincubator.com/fermentation-a-quick-guide-for-alt-protein-fans/. Accessed 16 June 2023

Pua A, Tang VCY, Goh RMV, Sun J, Lassabliere B, Liu SQ (2022) Ingredients, processing, and fermentation: addressing the organoleptic boundaries of plant-based dairy analogues. Foods 11. https://doi.org/10.3390/foods11060875

Silano V, Barat Baviera JM, Bolognesi C, Cocconcelli PS, Crebelli R, Gott DM, Grob K, Lambré C, Lampi E, Mengelers M, Mortensen A, Rivière G, Steffensen I-L, Tlustos C, van Loveren H, Vernis L, Zorn H, Aguilera J, Andryszkiewicz M, de Sousa RF, Liu Y, Di Piazza G, Rainieri S, Chesson A (2022) Safety evaluation of the food enzyme chymosin from the genetically modified Aspergillus niger strain DSM 29546. EFS2 20:e07465. https://doi.org/10.2903/j.efsa.2022.7465

Tuomisto HL, Scheelbeek PFD, Chalabi Z, Green R, Smith RD, Haines A, Dangour AD (2017) Effects of environmental change on population nutrition and health: a comprehensive framework with a focus on fruits and vegetables. Wellcome Open Res 2:21. https://doi.org/10.12688/wellcomeopenres.11190.2

[9] Perfect Day (2023); TurtleTree (2023).

TurtleTree (2023) Precision fermentation perfected: Fermentation 101: everything you ever wanted to know about precision fermentation, the innovative technology set to revolutionize alternative food. https://turtletree.com/precision-fermentation-perfected-fermentation-101/. Accessed 16 June 2023

Chapter 4
Pre-market Authorisation

4.1 Intro: Applicable Legal Frameworks

There is no specific legal framework exclusively dedicated to precision fermentation products intended for use as food or in food production. However, various laws may apply to these innovative products within the food sector, depending on factors such as their categorisation, composition, and production method. Alongside the GFL, there are more stringent regulations that could be relevant, such as the NFR and the GMFR. Both of these laws mandate pre-market authorisation for products falling under their scope. Additionally, if genetically modified microorganisms (GMMs) are used in the production process, producers must comply with the transpositions at the Member State level of the Directive on the contained use of genetically modified microorganisms.[1] During the production process and market placement, which includes transportation (for example to trade fairs or tastings) and retail, strict adherence to EU rules concerning hygiene and controls[2] is mandatory.

The utilisation of microorganisms and innovative technologies in food production can present regulatory challenges. In the subsequent chapters, we will delve into the legal framework and the EU requirements necessary for the lawful placement of precision fermentation-derived dairy alternatives on the market.

Within the context of EU food law, the concept of "**placing on the market**" is broad[3] and comprises several stages and actions. It is stipulated as "the holding of

[1] Directive 2009/41/EC (2009).

[2] Regulation (EC) No 852/2004 (2004); Regulation (EU) 2017/625 (2017).

[3] Simpson (2020).

F. Ronchetti et al., *The Regulatory Landscape in the EU for Dairy Products Derived from Precision Fermentation*, SpringerBriefs in Law,
https://doi.org/10.1007/978-3-031-49692-9_4

food or feed for the purpose of sale, including offering for sale or any other form of transfer, whether free of charge or not, and the sale, distribution, and other forms of transfer themselves".[4]

Unlike other product legislation, "placing on the market" of food items does not only include the offering to the final consumer[5] or requiring a supply in terms of an "offer or agreement for transfer of rights."[6] The mere act of holding it with that specific purpose is enough to fulfil the requirements of this definition.[7] Hence, any forms of offering and transfer is also included, making it e.g. unlawful to move the product fo example to trade fairs or organise tastings before pre-market authorisation is granted. These circumstances pose an additional hurdle for the development of innovative foods, as consumer acceptance is crucial to predict whether a food will be successful after market introduction, but cannot be tested and validated in advance. According to some of our interviewees, the absence of reliable consumer acceptance data is also likely to negatively impact investors' interest in the product and its development. Investors are hesitant to invest in such products without comprehensive testing and validation. Testing and validation provide essential evidence of the product's safety, quality, and efficacy, instilling confidence in both consumers and investors alike. These consequences have been acknowledged also in a report commissioned by the UK Food Safety Authority.[8] One key takeaway is the necessity of a defined legal framework enabling food tastings, such as allowing taste tests to be conducted under strict conditions before the product is granted pre-market approval. Implementing such measures could foster a regulatory environment that prioritises business friendliness and innovation.

4.2 Using Microorganisms for Food Production

The precision fermentation process consists of deploying microorganisms to produce specific compounds. Depending on whether or not these microorganisms are genetically engineered, they may be classified as genetically modified microorganisms (GMMs). Notably, the EU developed rules regarding the contained use of GMMs and requires prior notification and/or authorisation of the premises used to carry out their utilisation. For food produced using non-genetically engineered microorganisms, additional safety assessments evaluating the risks posed by the chosen biological agents are also often required.[9] There are, however, some microbial species for which harmlessness to human and animal health and to the

[4] As per Article 3(8) GFL.

[5] van der Meulen (2018).

[6] Department for Exiting the European Union (2017), p. 5.

[7] Simpson (2020).

[8] FSA from Deloitte LLP (2023).

[9] This is the case for novel foods—see Sect. 4.3.1.2.

environment is pre-emptively assumed based on the current state of knowledge. Using species that enjoy this status in food production can be of advantage, as it simplifies the safety assessment and approval procedure of the final product.

Hence, even during the development of a new product and/or before starting production, it is recommended to consider the rules applicable to the contained use of GMMs and the possible deployment of microorganisms that have been granted the Qualified Presumption of Safety (QPS) status.

4.2.1 GMMs and Their Contained Use

In the EU, a directive on the contained use of GMMs[10] came into force in May 2009 and defines GMMs as microorganisms "in which the genetic material has been altered in a way that does not occur naturally by mating and/or natural recombination."[11] The definition is complemented by two lists of techniques provided in Annex I to the directive:

- Part A encompasses the techniques regarded as inducing "genetic modification" and highlights, among other things, the following: a) recombinant nucleic acid techniques involve creating new combinations of genetic material by the insertion into any vector system of nucleic acid molecules produced outside of it and their incorporation into a host organism. In this context, the host organism may not naturally contain these genetic materials, but it is capable of continuing their propagation. b) The techniques in question also involve directly introducing heritable material prepared outside the micro-organism into a micro-organism. This includes methods such as micro-injection, macro-injection, and micro-encapsulation. Such processes facilitate the transfer of genetic material into the micro-organism from external sources, altering its genetic composition and characteristics.[12]
- Part B outlines techniques that are not considered to lead to genetic modification. These techniques include a) in vitro fertilisation; b) natural processes such as: conjugation, transduction, transformation; and c) polyploidy induction, provided that no use of recombinant-nucleic acid molecules or GMMs made by techniques/methods other than the techniques/methods excluded by Part A is involved.[13]

When evaluating the applicability of the GMM Directive, it is essential to recognize that the definition of microorganisms encompasses animal and plant cells in culture,

---

[10] Directive 2009/41/EC (2009) (hereinafter GMM directive).

[11] GMM Directive. Article 2(b).

[12] GMM Directive. Annex I Part A.

[13] GMM Directive. Annex I Part B.

in addition to viruses, viroids, and any other biological entity (cellular or non-cellular) capable if replication or of transferring genetic material.[14]

If precision fermentation involves the use of one or more GMMs, the requirements outlined in the GMM Directive must be met.[15] The primary objective of this Directive is to safeguard human health and the environment from potential risks associated with GMMs by developing measures for their so-called “contained use”. This term encompasses “any activity in which micro-organisms are genetically modified or in which such GMMs are cultured, stored, transported, destroyed, disposed of or used in any other way, and for which specific containment measures are used to limit their contact with, and to provide a high level of safety for, the general population and the environment.”[16]

Before utilising a GMM, the natural or legal person responsible for its contained use (the so-called user[17]) must conduct an assessment of the relative risk to human health and the environment. This evaluation includes considerations for waste and effluent disposal. Following the assessment, the GMM is then assigned to one of the following classes based on the level of risk identified:[18]

- Class 1: for activities with no or negligible risk.
- Class 2: for activities with low risk.
- Class 3: for activities with moderate risk.
- Class 4: for activities with high risk.

When a premise is intended to be used for the first time for the handling of a GMM in a contained environment, the user is required to submit a notification to the competent authority of their respective EU country.[19] The notification must include information such as the user’s name, the training and qualifications of the persons responsible for supervision and safety address, and a general description of the premises of the nature of the work which will be undertaken.[20]

For class 1 contained use of GMMs the notification procedure is sufficient, allowing the user to proceed with their activities without further delay. For class 2 contained use, an extended notification that includes additional information is necessary.[21] If the premises have been previously notified for carrying out class 2 or higher activities, the user can proceed immediately after notifying the competent national authority. In the absence of previous notification, the user must wait for a

[14] Acc. to GMM Directive. Article 2(a).

[15] Exceptions thereto are in place for contained uses involving only types of GMMs meeting the criteria which establish their safety for human health and the environment, acc. to GMM Directive. Article 3(1)(b). These (general as well as specific) criteria are listed in GMM Directive. Annex II, Part B.

[16] GMM Directive. Article 2(c).

[17] GMM Directive. Article 2(e).

[18] GMM Directive. Article 4 (1–3 and 5).

[19] GMM Directive. Article 6.

[20] A conclusive list of all required information is listed in GMM Directive. Annex V Part A.

[21] GMM Directive. Annex V Part A.

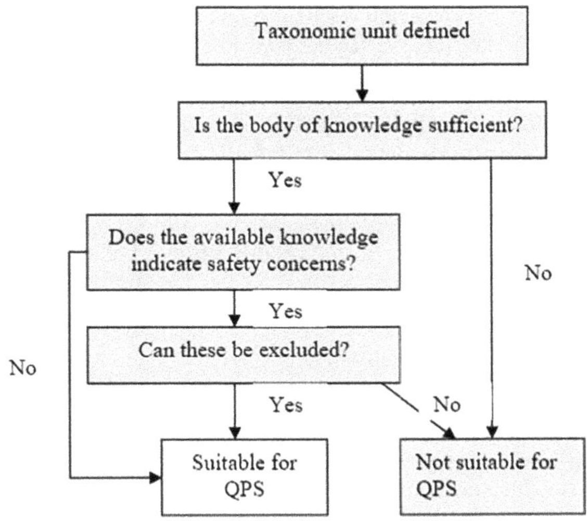

Fig. 4.1 A generalised scheme for assessing the suitability for QPS status of microorganisms (European Food Safety Authority 2005)

positive response from the competent national authority. If, after 45 days from the submission of the notification, no refusal has been issued by the competent authority, the user is permitted to proceed with the activities as notified.[22] Class 3 and class 4 contained uses require prior authorisation of the competent authority. The authority must communicate its decision in writing. The notification for class 3 and class 4 contained uses must contain all the information outlined in Annex V Part C of the GMM Directive. The decision of the competent authority regarding the authorisation must be issued within 45 or 90 days after the submission, depending on whether it is the first or a subsequent notification for the premise.[23]

4.2.2 *Qualified Presumption of Safety*

When using microorganisms in food production, it is recommended to utilise biological agents that have been granted a Qualified Presumption of Safety (QPS). The QPS concept was introduced by EFSA in relation to the safety assessment approach of microorganisms used in food or feed, aiming to streamline the safety evaluation of products seeking market authorisation. The QPS status is granted to strains that are already well-known and thoroughly analysed, with several safety aspects having been assessed. This status, based on reasonable evidence and careful research, exempts selected groups of microorganisms from the requirement for further safety assessments.[24]

[22] GMM Directive. Article 8.

[23] GMM Directive. Article 9.

[24] European Food Safety Authority (2005).

The EFSA Panel on Biological Hazards (BIOHAZ Panel) establishes whether a biological agent is granted a QPS status. The BIOHAZ Panel pre-assesses safety concerns for humans, animals and the environment. The taxonomic identity of the microorganism, the related body of knowledge and potential safety concerns (*inter alia* the antimicrobial resistance) are inspected. Figure 4.1 shows the decision tree at the core of the QPS assessment.[25]

Once a microorganism receives the QPS status, it is included in a positive list together with its required qualifications. The QPS positive list is updated every three years by the BIOHAZ Panel:[26] the latest Update was published in January 2023.[27] As of now, 111 species (including bacteria, protists/algae, viruses, yeasts) were granted the QPS status.[28]

As the QPS assessment is conducted independently of the mandatory safety assessments for products requiring pre-market authorisation, the fact that a food contains or is produced with a QPS microorganism does not automatically lead to its authorisation for market placement. It may, however, simplify and shorten the approval process. Non-QPS strains may also be used in food production. If this is the case, however, they must undergo a full safety assessment: more data supporting safety must be generated, which leads to higher costs in terms of money and time.

4.3 Placing Food on the EU Market

Products falling under the definition of "food" or "foodstuff"[29] are generally allowed on the EU market without prior authorisation or notice, as long as they

- have been consumed on the EU market before 15. May 1997,
- are not unsafe (i.e. neither injurious to health nor unfit for human consumption),[30] and
- comply with the requirements related to traceability and labelling set out in the GFL[31] and other legislation such as the FIR.

Certain product categories are deemed inherently risky by the legislator, and as a result, they must undergo pre-market authorisation to ensure their safety through

[25] Figure 1 borrowed from: European Food Safety Authority (2005).

[26] EFSA Panel on Biological Hazards.

[27] Koutsoumanis et al. (2023).

[28] Zenodo (2023).

[29] GFL. Article 2: the terms 'food' and 'foodstuff' describe "any substance or product, whether processed, partially processed or unprocessed, intended to be, or reasonably expected to be ingested by humans".

[30] GFL. Article 14.

[31] GFL. Articles 18, 16; national rules GFL. Article 14(9).

scientific assessments.[32] These categories include novel foods, "genetically modified foods", and various food improvement agents such as additives, food enzymes and flavourings.

Hence, to comprehend the necessary requirements for lawfully marketing a new food product in the EU, it is imperative to assess the categories to which the product itself and its components belong. Adherence to the appropriate regulations and completion of the necessary authorisation processes within these categories are crucial steps in ensuring the product's lawful entry into the EU market.

In the case of dairy product alternatives containing casein obtained through precision fermentation (i.e., where casein is used as a food component or ingredient), the categories "additives," "food enzymes," and "flavourings" can be excluded a priori. However, categorising these products as either novel or genetically modified (GM) foods is not equally straightforward. During the initiation of this project, we conducted stakeholder interviews and received a wide range of responses to the classification question. This diversity of opinions highlighted the uncertainties among food business operators (FBOs) and experts in the field when it comes to determining the appropriate legal framework for these products. These uncertainties are linked to various variables characterizing the production process, as well as the challenges of navigating a regulatory environment that is often perceived as outdated or inadequate for accommodating certain food innovations.[33] The absence of legal certainty regarding market authorisation and labelling requirements may impede innovation-driving companies from planning their steps reliably and, as a result, deter potential investors.[34]

4.3.1 Product Categorisation

This section presents the steps that the FBO should undertake to correctly classify their precision-fermented dairy product under EU food law. An accurate determination of the foodstuff's category enables the timely recognition of the legal requirements the product must comply with to be lawfully marketed in the EU. Once the categorisation is finalised, if necessary, FBOs can proceed to create a dossier for the appropriate pre-market application procedure.

From our introductory interviews, a widespread uncertainty about the product's classification among stakeholders emerged, especially when participants were asked whether precision-fermented dairy alternatives would be considered GM foods or novel foods. The responsibility of choosing the appropriate framework and applying for the correct approval procedure lies with the FBO.[35]

[32] European Commission (2000).

[33] FSA from Deloitte LLP (2023); Holle (2018); Monaco and Purnhagen (2022).

[34] Holle (2018); Molitorisová et al. (2021).

[35] See for example NFR. Article 4(1).

FBOs are allowed to submit an application dossier for different regulatory frameworks. However, it is essential to consider that the prerequisites, requirements, and procedures can vary significantly between different regulatory categories. Pursuing multiple applications can be time-consuming and financially burdensome. Therefore, it is highly recommended to identify the correct product category and applicable legal framework at an early stage.

Consequently, a thorough analysis of the GMFR and NFR is indispensable. This proactive approach will enable FBOs to focus their efforts and resources on the most suitable and efficient regulatory pathway, ensuring a smoother and more successful market authorisation process for their food products. The first question that needs to be addressed is whether GMMs are utilised in the production process.

4.3.1.1 Precision Fermentation Employing GMMs

There is no specific legal framework explicitly designed to address precision fermented foodstuffs. In situations where food products contain genetically modified microorganisms (GMMs) or are produced using GMMs, and these GMMs and their products do not fall under the legal definitions of food additives, enzymes, or flavourings, they may be subject to either the Genetically Modified Food Regulation (GMFR) or the Novel Food Regulation (NFR).

Food produced "from" a GMO vs. food produced "with" a GMO: The GMFR covers "(a) GMOs for food use; (b) food containing or consisting of GMOs; (c) food produced from or containing ingredients produced from GMOs."[36] However, the GMFR does not explicitly specify whether precision fermentation products are included in its scope. In fact, recital 16 GMFR stipulates:

> **This Regulation should cover food and feed produced 'from' a GMO but not food and feed 'with' a GMO.** The determining criterion is whether or not material derived from the genetically modified source material is present in the food or in the feed. Processing aids which are only used during the food or feed production process are not covered by the definition of food or feed and, therefore, are not included in the scope of this Regulation. Nor are food and feed which are manufactured with the help of a genetically modified processing aid included in the scope of this Regulation.[37]

Recital 16 GMFR further clarifies that

> products obtained from animals fed with genetically modified feed or treated with genetically modified medicinal products will be subject neither to the authorisation requirements nor to the labelling requirements referred to in this Regulation.

This clarification pertains to the distinction between different scenarios involving genetically modified (GM) products. For instance, milk obtained from cows that have been fed with GM feed is not considered as a GM product, whereas soybean oil produced directly from GM soybeans is considered a GM food. The question

[36] GMFR. Article 3(1).

[37] GMFR. Recital 16.

arises, whether the transfer of material derived from the GM source mentioned in Recital 16 only refers to horizontal transfer of genetic material with reproductive ability, or if it includes any detectable trace of the employed GMO, regardless of its size, integrity and potential to reproduce.[38]

Additionally, according to Recital 16 GMFR sentence 3, foodstuffs manufactured with the help of a genetically modified processing aid are excluded from the scope of the GMFR. Processing aids are defined in Commission Regulation (EC) No. 1333/2008 on food additives[39] as

> any substance which:
>
> (i) is not consumed as a food by itself;
> (ii) is intentionally used in the processing of raw materials, foods or their ingredients, to fulfil a certain technological purpose during treatment or processing; and
> (iii) may result in the unintentional but technically unavoidable presence in the final product of residues of the substance or its derivatives provided they do not present any health risk and do not have any technological effect on the final product.[40]

Due to the significant number of questions and uncertainties regarding the practical application of the GMFR the Commission issued a report on its implementation in 2006.[41] In this report, sections 10.2 and 10.3 directly address GMM-fermentation derived products and provide clarifications on their status. The report affirms that the determination of such product's status must be assessed in the light of the recital 16 GMFR. It also specifies that GMMs used as processing aids are excluded from the scope of the GMRF. However, it emphasises that *"when the GMMs are not removed during the production process, they are not used as processing aids."*[42] Thus, the respective authorisation must be sought. Based on the statements provided, it seems that the critical factor in determining whether a GMM is used as processing aid is the removal of the GMM during the production process. If the GMMs are removed and do not remain in the final product, they could be considered as processing aids and be excluded from the scope of the GMFR. During the production of precision fermented dairy alternatives, the downstream processing involves the purification of casein, and the GMMs used for its synthetisation are removed therefrom. Initially, one might expect that GMMs used in precision fermentation would fall under the definition of processing aids, and, therefore, be excluded from the scope of the GMFR. However, the situation is more intricate, and a more stringent interpretation is necessary, as indicated below.

In October 2016 the DG SANTE of the EC sent a letter to EFSA concerning "Fermentation products obtained from GMMs."[43] In the letter, DG SANTE emphasised that the determining factor in establishing the regulatory context for various

---

[38] Dederer (2021).

[39] Regulation (EC) No 1333/2008 (2008).

[40] Regulation (EC) No 1333/2008 (2008). Article 3(2)(b).

[41] European Commission (2006).

[42] European Commission (2006). Section 10.2 and 10.3.

[43] van Goethem (2016).

sectoral laws related to products obtained via GMM precision fermentation lies in whether a food is produced "from" or "with" a GMO. In light of Recital 16 GMFR, this criterion must be interpreted based on **the presence or absence of recombinant DNA (rDNA[44]) in the obtained material**. This statement implies the following:

if in the final product recombinant rDNA is detected, it considered to have been produced from a GMO (i.e. it is a GMO) and is subject to pre-market authorisation required in Article 4(2) GMFR.

If no rDNA is detected in the final product, it is considered to have been produced with a GMO and thus falls out of the scope of the GMFR, and is subject to the appropriate sectoral legislation.

In practice, a GMM employed in precision fermentation is classified as a processing aid and excluded from the scope of the GMFR only if it is completely removed during the production process. This application contrasts with the definition of processing aid, which allows for the possibility of residues or derivatives of the substance to be unintentionally present in the final product, as long as they are safe and do not have further technological effect.

Critique to This Approach

The zero-tolerance approach towards detecting recombinant rDNA in GMM fermented products has faced significant criticism in literature. Under this approach, any detection of even the smallest amount of rDNA with any method in a fermentation product leads to the applicability of the GMFR. However, avoiding the presence of such extremely low traces of rDNA is both technologically and economically unfeasible.[45] Scholars propose an alternative solution by establishing a *de minimis* threshold, which involves adopting a detection threshold of 10 ng DNA/g or mL product. Below this threshold, it is presumed that there is no or only insignificant "presence of DNA from the production strain".[46] This threshold has already been suggested in two EFSA documents concerning the characterization of GMMs in feed additives and food enzymes.[47]

The letter from DG SANTE recognises that determining such a threshold is linked to three primary technical challenges related to the actual detection of rDNA in the final products. These main limitations include:

[44] rDNA is a compound DNA molecule formed through genetic recombination. In this specific case (precision fermented casein), it refers to the genetically engineered plasmid DNA of the host organism, that contains the inserted gene coding for casein synthesis.

[45] Dederer (2021).

[46] Silano et al. (2019).

[47] Rychen et al. (2018); Silano et al. (2019).

- The absence of standardised rDNA detection methodology;
- Technological advantages in DNA detection methodologies lead to extremely high test sensitivity;
- The absence of an established threshold/cut-off value of rDNA *"to draw a line between 'real' and 'fortuitous' presence of rDNA"*[48] in GMM fermentation products.

Authorities are facing substantial difficulties in deciding on the regulatory status of GMM fermented products and require a *"harmonised, scientifically sound, and pragmatic solution to be applied consistently across all legislations."*[49] In its letter, DG SANTE acknowledges the importance of addressing these challenges and expresses support for the guidelines and thresholds proactively developed by the FEEDAP Panel.[50] However, before moving forward with regulatory implementation, the EC requires the formal endorsement of these guidelines by other relevant EFSA panels, including the GMO, ANS, CEF.[51] This step has not yet been officially accomplished. As of now, this endorsement process has not yet been officially completed.

GMMs as processing aids and the categorisation as novel foods: GMMs utilised in precision fermented foods are considered processing aids only if they are completely removed from the final product, or at least to the extent that no rDNA is detectable. Processing aids are defined by Regulation (EC) No. 1333/2008 on food additives, but excluded from the scope and there is no specific European legislation governing processing aids. While individual Member States might maintain specific regulations concerning processing aids, there is currently no mandatory labeling requirement for these aids at the EU level.

Precision fermented foods that employ GMMs as processing aids typically fall within the category of novel foods. Novel foods encompass a diverse group of foods that were *"not used for human consumption to a significant degree within the Union before 15 May 1997, irrespective of the dates of accession of Member States to the Union"*,[52] and that fall under at least one of ten specific categories, *inter alia*:

"(ii) food consisting of, isolated from or produced from microorganisms, fungi or algae;

(vi) food consisting of, isolated from or produced from cell culture or tissue culture derived from animals, plants, microorganisms, fungi or algae;

---

[48] van Goethem (2016).

[49] van Goethem (2016).

[50] EFSA Panel on Additives and Products or Substances used in Animal Feed.

[51] Scientific Panel on Genetically Modified Organisms (GMO); Scientific Panel on Food Additives and Nutrient Sources added to Food (ANS); Panel of Food Contact Material, Enzymes, Flavourings and Processing Aids Panel (CEF).

[52] NFR. Article 3(2)(a).

(vii) food resulting from a production process not used for food production within the Union before 15 May 1997, which gives rise to significant changes in the composition or structure of a food, affecting its nutritional value, metabolism or level of undesirable substances".[53]

Both the novelty of the process and the direct use of microorganisms in the production of precision fermented dairy alternatives are valid reasons for the applicability of the NFR to this specific product category. If FBOs are uncertain about whether their food product falls under the definition of a novel food, they have the option to seek guidance from the competent authority of the Member State where they intend to place their product on the EU market.[54] The procedural steps for this consultation process are outlined in Commission Implementing Regulation (EU) 2018/456.[55] When submitting a consultation request, the FBO must provide the necessary information for the product's classification (including a technical dossier and supporting documentation). Currently, no food product derived from precision fermentation using GMMs has undergone categorisation through the consultation process. However, in May 2019 the Portuguese company Setenta e Três Mil e Cem Lda sought consultation regarding L-fucose, which was produced by non-GMO microbial fermentation.[56] The recipient Member State authority confirmed the novel food status of the product under novel food category (ii) of Article 3 (2)(a) NFR.[57]

Current developments: In recent years, the EC has granted approval for several novel foods produced through GMM-based precision fermentation, included them in the Union List of novel foods and consequently allowing them to be placed on the market.[58] The specific novel foods that have been added to the Union List as the latest approved additions are detailed below:

- L-Alanyl-L-Glutamine (a food supplement obtained via fermentation with a genetically modified strain of *Escherichia coli*) and the Ice Structuring Protein type III HPLC 12 (produced by submerged fermentation of a genetically-modified strain of food-grade *Saccharomyces cerevisiae*[59]) were approved under the old novel food regime.[60]

---

[53] NFR. Article 3(2)(a).

[54] NFR. Article 4(2).

[55] Commission Implementing Regulation (EU) 2018/456 (2018).

[56] Novel Food Unit of the Medicines Evaluation Board Agency (2019).

[57] Novel Food Unit of the Medicines Evaluation Board Agency (2019).; L-fucose is thereby suggested to fall additionally under novel food category (i) "*food with a new or intentionally modified molecular structure, where that structure was not used as, or in, a food within the Union before 15 May 1997*", but this would not be the case for precision fermented casein that is bioidentical to conventionally produced casein.

[58] Commission Implementing Regulation (EU) 2017/2470 (2017).

[59] Interestingly, in the Union list of novel foods, one of the specifications for this product in the column "Conditions under which the novel food may be used" is "DNA: Not detectable".

[60] 2009/344/EC: Commission Decision (2009); Regulation (EC) No 258/97 (1997).

2'-Fucosyllactose (microbial source)	**Source:** Genetically modified strain of *Escherichia coli* K-12	**Source:** Genetically modified strain of *Escherichia coli* BL-21	**Source**: Genetically modified strain of *Corynebacterium glutamicum* ATCC 13032

Fig. 4.2 Microbial sources of approved 2′-Fucosyllactose

- Lacto-N-tetraose (LNT) produced from microbial fermentation of with a genetically modified strain of *Escherichia coli,* strain K12 DH1 in 2020[61] and of derivative strains of *Escherichia coli* BL21(DE3) in 2023,[62] both under data protection in accordance with Article 26 NFR.
- 3′-Sialyllactose (3′-SL): sodium salt obtained by microbial fermentation with the genetically modified strain of *Escherichia coli,* strain K12 DH1, approved in 2021 and data protected until 2026.[63]
- 2′-Fucosyllactose obtained through microbial fermentation of the strains listed in Fig. 4.2 was authorised for three different applicants between 2016 and 2023.[64]
- 2′-Fucosyllactose/Difucosyllactose mixture ('2′-FL/DFL') obtained by microbial fermentation with a genetically modified strain of Escherichia coli strain K12 DH1 was approved in 2019.[65] Proprietary data are protected until 19.12.2024.
- 3-Fucosyllactose (3-FL) obtained by microbial fermentation with a genetically modified strain of Escherichia coli, strain K12 MG1655, was authorised in December 2021.[66] The same product, however produced by a derivative strain of *Escherichia coli* BL21(DE3), gained approval in early 2023.[67] In both cases, the applicants obtained the five years of protection for their proprietary data.

4.3.1.2 Precision Fermentation Without Employment of GMMs

If the production process of precision fermented foodstuffs does not involve the use of GMMs, these products may will fall within the scope of the NFR, especially under the novel food categories (ii), (vi) and/or (vii) of Article 3 (2)(a) NFR.

FBOs who are uncertain about whether their product qualifies as a novel food can request a consultation process according to Article 4 NFR with the competent authority of the Member State where they intend to place their product on the market.[68] The outcome of this consultation process has, however, no legally binding value and is hence only of limited use for FBOs.

---

[61] Commission Implementing Regulation (EU) 2020/484 (2020).

[62] Commission Implementing Regulation (EU) 2023/7 (2023).

[63] Commission Implementing Regulation (EU) 2021/96 (2021).

[64] Commission Implementing Regulation (EU) 2023/859 (2023).

[65] Commission implementing Regulation (EU) 2019/1979 (2019).

[66] Commission Implementing Regulation (EU) 2021/2029 (2021).

[67] Commission Implementing Regulation (EU) 2023/52 (2023).

[68] See the case of L-fucose produced by (non-GMO) microbial fermentation by the Portuguese company Setenta e Três Mil e Cem Lda regarding the request for consultation in footnote 94.

4.3.2 Pre-market Authorisation Under the GMFR

Foodstuffs that fall within the scope of the GMFR are not permitted to be placed on the market unless they have received authorisation.[69] The general requirements for pre-market authorisation are that such foods do not

(a) have adverse effects on human health, animal health or the environment;
(b) mislead the consumer;
(c) differ from the food which it is intended to replace to such an extent that its normal consumption would be nutritionally disadvantageous for the consumer.[70]

The steps characterising the pre-market approval procedure are specified in Chapter II Section 1 GMFR. First, the applicant must submit the application dossier to the national competent authority of a Member State, which, without delay, makes all submission documentation available to EFSA.

Pursuant to GMFR Article 5(3), the application must contain:

(a) the name and the address of the applicant;
(b) the designation of the food, and its specification, including the transformation event(s) used;
(c) where applicable, the information to be provided for the purpose of complying with (…) the Cartagena Protocol on Biosafety to the Convention on Biological Diversity;
(d) where applicable, a detailed description of the method of production and manufacturing;
(e) a copy of the studies, including, where available, independent, peer-reviewed studies, which have been carried out and any other material which is available to demonstrate that the food complies with the (general requirements);
(f) either an analysis, supported by appropriate information and data, showing that the characteristics of the food are not different from those of its conventional counterpart (…);
(g) either a reasoned statement that the food does not give rise to ethical or religious concerns, or a proposal for labelling it (in this respect);
(h) where appropriate, the conditions for placing on the market the food or foods produced from it, including specific conditions for use and handling;
(i) methods for detection, sampling (including references to existing official or standardised sampling methods) and identification of the transformation event and, where applicable, for the detection and identification of the transformation event in the food and/or in foods produced from it;
(j) samples of the food and their control samples, and information as to the place where the reference material can be accessed;
(k) where appropriate, a proposal for post-market monitoring regarding use of the food for human consumption;
(l) an identification of the parts of the application and any other supplementary information that the applicant requests to be treated as confidential, accompanied by verifiable justification (…)
(m) a summary of the dossier in a standardised form.

[69] GMFR. Article 4(2).

[70] GMFR. Article 4(1).

Additional requirements are specified in Commission Implementing Regulation (EU) No. 503/2013 of 3 April 2013 on applications for authorisation of genetically modified food and feed.[71]

After receiving the submission documents, EFSA informs the other Member States and the European Commission and makes a summary of the dossier publicly available.[72] Within six months from the receipt of a valid application, EFSA is required to issue a scientific opinion regarding the safety of the product. In some cases, EFSA may request additional information from the applicant to aid in the assessment, which can extend the time limit for EFSA's risk assessment. This described procedure, which includes a public consultation phase, extends the time limit set for the development of EFSA's risk assessment.[73] After EFSA completes its risk assessment, the final opinion is made public, followed by a 30 day long public consultation phase. After the public consultation, the EC, assisted by the Standing Committee on the Food Chain and Animal Health, may grant approval for the product after a successful completion of the committee procedure. This approval process is subject to a three-months-time frame from the date of EFSA's assessment. If the EC's opinion differs from that of EFSA, the EC is required to provide a comprehensive explanation of its reasoning for deviating from EFSA's assessment.[74] The EC has established a Community register of GM food and feed in accordance with Article 28 GMFR. This register is included in the EU Register of authorised GMOs and can be accessed at the following link: https://webgate.ec.europa.eu/dyna2/gm-register/. As of the current date, there have been no precision fermented products authorised under the GMFR.

Despite the time frames for processing application specified by law, it has been highlighted by several interviewees that these time frames are often not adhered to. The evaluation process may experience delays due to various reasons. For instance, EFSA may request additional information from the applicant, leading to a "clock-stop" where the time frame is paused until the requested data is provided. Additionally, the risk assessment capacities of EFSA can be limited or saturated, which may impact the efficiency of the evaluation process. The political step of the authorisation process can also be time-consuming and sometimes subject to postponement to a later session. Such delays may result in the process taking longer than initially anticipated, causing uncertainties for applicants seeking authorisation under the GMFR.

[71] Commission Implementing Regulation (EU) No 503/2013 (2013).

[72] GMFR. Article 5.

[73] GMFR. Article 6(1,2).

[74] GMFR. Article 5(7) and Article 7(1).

Under the GMFR authorisations for GM food and feed are granted for 10-years. To renew the authorisation, the current authorisation-holder must submit a renewal-application to the EC at the latest one year before the current authorisation's expiry date.[75] In addition to the 10-year authorisation, the GMFR provides applicants with a ten-year protection period for the scientific data and the other information included in the application dossier.[76]

GMO authorisations for food - Application guidance: The application process for the pre-market authorisations of foods regulated by the GMFR requires the use of a dedicated electronic tool: the E-Submission Food Chain Platform.[77] This platform allows applicants to submit their applications online and monitor the procedure until a final outcome is reached.

Over the years, EFSA published several guidance documents to support applicants through the application process. These guidances include "Administrative guidance on the submission of applications for renewal of authorisation of genetically modified food and feed under Articles 11 and 23 of Regulation (EC) No. 1829/2003"[78] and "Administrative guidance for the preparation of renewal applications on genetically modified food and feed."[79]

Guidance document on the risk assessment of GMMs in food: To aid applicants in presenting their applications in accordance with the GMFR and to provide guidance on the risk assessment of GMMs used in food production, EFSA published an updated Guidance on the risk assessment of genetically modified microorganisms and their products intended for food and feed use.[80] According to this guidance, the risk characterisation of GMMs should be carried out on a case-by-case basis, taking into account various criteria. These criteria include the category of the GMM and/or product, the type of product obtained from it, the genetic modification, the production process and on the expected use of the derived food.[81] Given the diversity of the GMMs and their products used in the food and feed sector, Chapter II offers an exemplary recommended characterisation of the products. This characterisation aims to facilitate a more harmonised risk assessment and classification process.

Category 1: Chemically defined purified compounds and their mixtures produced with GMMs in which both GMMs and newly introduced genes have been removed (e.g. amino acids, vitamins);

Category 2: Complex products produced a) with GMMs in which both GMMs and newly introduced genes have been removed (e.g. most enzyme preparations); or b) from GMMs in which both GMMs and newly introduced genes are no longer present (e.g. cell extracts);

---

[75] GMFR. Article 5(7), Article 7(1), and Article 11(1).

[76] GMFR. Article 31.

[77] European Commission (2023).

[78] European Food Safety Authority (2019).

[79] European Food Safety Authority (2021b).

[80] European Food Safety Authority (2011).

[81] European Food Safety Authority (2011), p. 32.

Category 3: Products produced from GMMs in which GMMs capable of replication or of transferring newly introduced genes are not present; but in which newly introduced genes are still present (e.g. heat-inactivated starter cultures);
Category 4: Products consisting of or containing GMMs capable of replication or of transferring newly introduced genes (e.g. live starter cultures).

When seeking authorisation for dairy alternatives produced with precision fermented casein, which involves the use of GMMs, BOs n, must consider the requirements applicable to their product category.

4.3.3 Pre-market Authorisation Under the NFR

Similarly to the provisions of the GMFR, foods falling under the scope of the NFR must undergo a pre-market authorisation. The EC has to establish and update a Union list of authorised novel foods.[82] The general conditions for an inclusion in the Union list are as follows:

> (a) the food does not, on the basis of the scientific evidence available, pose a safety risk to human health;
> (b) the food's intended use does not mislead the consumer, especially when the food is intended to replace another food and there is a significant change in the nutritional value;
> (c) where the food is intended to replace another food, it does not differ from that food in such a way that its normal consumption would be nutritionally disadvantageous for the consumer.[83]

To request the inclusion of a new product on the Union list, the FBOs must follow the application procedure set out in Article 10 NFR.[84] This procedure entails submitting a comprehensive dossier that covers all relevant information about the single novel food they intend to introduce to the EU market.

> The application for an authorisation shall include:
>
> (a) the name and address of the applicant;
> (b) the name and description of the novel food;
> (c) the description of the production process(es);
> (d) the detailed composition of the novel food;
> (e) scientific evidence demonstrating that the novel food does not pose a safety risk to human health;
> (f) where appropriate, the analysis method(s);
> (g) a proposal for the conditions of intended use and for specific labelling requirements which do not mislead the consumer or a verifiable justification why those elements are not necessary.[85]

---

[82] Annex to Commission Implementing Regulation (EU) 2017/2470 (2017).

[83] NFR. Article 7.

[84] The details of this procedure are set out in Commission Implementing Regulation (EU) 2017/2469 (2017).

[85] NFR. Article 10(2).

When it comes to the safety assessments of novel foods, the burden of proof is lies the applicant. The applicant must provide robust and comprehensive scientific evidence demonstrating the product's harmlessness. This evidence may include numerous toxicity studies, animal models or human data, among others[86] The applicant is obliged to submit all available scientific data, whether proprietary, confidential or published, relevant to the safety of the novel food, regardless of whether they support or oppose their product.[87]

Once a valid application is submitted, EFSA conducts a thorough evaluation of the dossier to determine whether the said novel food is liable to have an effect on human health. Following EFSA's evaluation, the EC makes the final decision on whether to include the novel food in the Union list. This decision is made following a successful application of the committee procedure, with the assistance of the Standing Committee on Plants, Animals, Food and Feed.[88] The EC considers EFSA's opinion, the precautionary principle, other relevant EU provisions considered applicable and *"any other legitimate factors relevant to the application under consideration"*,[89] including economic or societal aspects or considerations related to tradition, ethics and the environment.[90] The entry of a novel food in the Union list includes not only the name of the authorised foodstuff and its composition specifications but also binding conditions of use and additional requirements, such as specific labelling requisites.[91] Once a product is included in the Union list, it can be lawfully placed on the EU market, provided that it is used in accordance with the specified conditions.[92]

Similar to GMO food pre-market authorisations, the time frame for novel food authorisation processes is defined by law but is often exceeded in practice.[93]

The data protection clause: According to Article 23 NFR, once EFSA receives a valid application, it is required to make the dossier, relevant supporting information, and any supplementary information provided by the applicant publicly available.[94] EFSA's scientific opinion must also be published. This is to ensure transparency and allow stakeholders and the public to access information about the novel food and its safety assessment. However, the NFR allows the applicant to request confidentiality for certain parts of the submitted information, and this confidentiality can be granted for a non-renewable period of five years.[95]

---

[86] European Food Safety Authority (2021a).

[87] Turck et al. (2016).

[88] NFR. Articles 12(1) and 30(1).

[89] NFR. Article 12(1).

[90] Volpato (2022).

[91] NFR. Article 9(3).

[92] NFR. Article 6.

[93] Holle (2018).

[94] Such information includes *inter alia* scientific outputs, scientific data and studies.

[95] NFR. Article 23 and Chapter V (Articles 26–28).

To request confidentiality, the applicant must provide a verifiable justification explaining why the disclosure of certain information could potentially harm its interests to a significant degree. This often applies to data related to proprietary scientific studies or sensitive business information. By granting confidentiality treatment, the NFR offers applicants the opportunity to protect their newly developed scientific data and other confidential information from being used by competitors for similar applications during the five-year period. This provision gives applicants a competitive advantage by safeguarding their research and development efforts and preventing others from benefiting from the data they provided in their application.[96] While the NFR typically grants a generic authorisation for a novel food, independent of the applicant who first obtained market placement permission for the product, the confidentiality clause allows applicants to retain a level of exclusivity for their data and protect their competitive edge for a limited period.

Overall, the confidentiality provision in the NFR strikes a balance between transparency and protection of commercial interests, ensuring that relevant information is accessible to the public while allowing applicants to safeguard sensitive data and maintain a competitive advantage in the market.

Novel food application guidance: The novel food application must also be submitted through the E-Submission Food Chain Platform.[97] In 2021, EFSA published the "Administrative guidance for the preparation of applications on novel foods pursuant to Article 10 of Regulation (EU) 2015/2283,"[98] which provides guidance and support to applicants regarding administrative matters. It outlines the procedure and the associated timelines for the handling of applications, the available support initiatives for applicants and the possibilities for interaction of the applicant with EFSA. EFSA published a new revised version of the "Guidance on the preparation and submission of an application for authorisation of a novel food in the context of Regulation (EU) 2015/2283"[99] in the same year. The guidance document serves to support applicants who are preparing a novel food application. It offers detailed information on the technical requirements and specific criteria related to the single novel food categories that must be met when submitting an application for novel food approval.

References

Dederer H (2021) rDNA traces in fermentation products using Genetically Modified Microorganisms (GMMs). StoffR 18:135–147. https://doi.org/10.21552/stoffr/2021/3/6

EFSA Panel on Dietetic Products, Nutrition and Allergies (2021) Guidance on the preparation and submission of an application for authorisation of a novel food in the context of Regulation (EU) 2015/2283 (Revision 1)2. EFS2 19:e06555. https://doi.org/10.2903/j.efsa.2021.6555

---

[96] Molitorisová et al. (2021).

[97] European Commission (2023).

[98] European Food Safety Authority (2021a).

[99] EFSA Panel on Dietetic Products, Nutrition and Allergies (2021).

European Commission (2000) Communication from the Commission on the precautionary principle. https://eur-lex.europa.eu/LexUriServ/LexUriServ.do?uri=COM:2000:0001:FIN:en:PDF Accessed 1 January 2024
European Commission (2006) Report from the Commission to the Council and the European Parliament on the implementation of Regulation (EC) No 1829/2003 of the European Parliament and of the Council on genetically modified food and feed. https://eur-lex.europa.eu/legal-content/EN/TXT/?uri=COM%3A2006%3A0626%3AFIN Accessed 1 January 2024
European Commission (2023) E -SUBMISSION: Food Chain platform. https://webgate.ec.europa.eu/esfc/#/applications. Accessed 11 Aug 2023
European Food Safety Authority (2005) Opinion of the Scientific Committee on a request from EFSA related to a generic approach to the safety assessment by EFSA of microorganisms used in food/feed and the production of food/feed additives. EFS2 3:226. https://doi.org/10.2903/j.efsa.2005.226
European Food Safety Authority (2011) Guidance on the risk assessment of genetically modified microorganisms and their products intended for food and feed use. EFS2 9:2193. https://doi.org/10.2903/j.efsa.2011.2193
European Food Safety Authority (2019) Administrative guidance on the submission of applications for renewal of authorisation of genetically modified food and feed under Articles 11 and 23 of Regulation (EC) No 1829/2003. EFS3 16. https://doi.org/10.2903/sp.efsa.2019.EN-1668
European Food Safety Authority (2021a) Administrative guidance for the preparation of applications on novel foods pursuant to Article 10 of Regulation (EU) 2015/2283. EFS3 18. https://doi.org/10.2903/sp.efsa.2021.EN-6488
European Food Safety Authority (2021b) Administrative guidance for the preparation of renewal applications on genetically modified food and feed. EFS3 18. https://doi.org/10.2903/sp.efsa.2021.EN-6474
FSA from Deloitte LLP (2023) Novel Foods Regulatory Framework Review: Executive Summary. https://www.food.gov.uk/print/pdf/node/17626. Accessed 7 Aug 2023
Goethem, B van (2016) Letter from the EC to EFSA, Subject: Fermentation products obtained from Genetically Modified Microorganisms (GMMs): Ref. Ares(2016)6016209 - 19/10/2016, SANTE/E3/TD/nn sante.ddg2.e.3(2016)6287877' (Brussels 16 October 2016)
Holle M (2018) Pre-market approval and its impact on food innovation: the novel foods example. In: Bremmers H, Purnhagen K (eds) Regulating and managing food safety in the EU. Springer International Publishing, Cham, pp 291–330
Koutsoumanis K, Allende A, Álvarez-Ordóñez A, Bolton D, Bover-Cid S, Chemaly M, de Cesare A, Hilbert F, Lindqvist R, Nauta M, Peixe L, Ru G, Simmons M, Skandamis P, Suffredini E, Cocconcelli PS, Fernández Escámez PS, Maradona MP, Querol A, Sijtsma L, Suarez JE, Sundh I, Vlak J, Barizzone F, Hempen M, Correia S, Herman L (2023) Update of the list of qualified presumption of safety (QPS) recommended microorganisms intentionally added to food or feed as notified to EFSA. EFS2 21. https://doi.org/10.2903/j.efsa.2023.7747
Molitorisová A, Monaco A, Purnhagen KP (2021) An analysis of the regulatory framework applicable to products obtained from mushroom and mycelium. SSRN J. https://doi.org/10.2139/ssrn.3955899
Monaco A, Purnhagen KP (2022) Risk triggers as innovation triggers? Risk analysis and innovation's promotion under the Novel Food Regulation. EFFL 17:219–227
Novel Food Unit of the Medicines Evaluation Board Agency (2019) Consultation request to determine the status of L-fucose (produced by microbial fermentation), pursuant to Article 4(2) of Regulation (EU) 2015/2283 of the European Parliament and of the Council of 25 November 2015 on novel foods. https://food.ec.europa.eu/system/files/2019-09/novel-food_consult-status_l-fucose.pdf. Accessed 16 June 2023
Rychen G, Aquilina G, Azimonti G, Bampidis V, Bastos ML, Bories G, Chesson A, Cocconcelli PS, Flachowsky G, Gropp J, Kolar B, Kouba M, López-Alonso M, López Puente S, Mantovani A, Mayo B, Ramos F, Saarela M, Villa RE, Wallace RJ, Wester P, Glandorf B, Herman L, Kärenlampi S, Aguilera J, Anguita M, Brozzi R, Galobart J (2018) Guidance on the characteri-

sation of microorganisms used as feed additives or as production organisms. EFS2 16:e05206. https://doi.org/10.2903/j.efsa.2018.5206

Silano V, Barat Baviera JM, Bolognesi C, Brüschweiler BJ, Cocconcelli PS, Crebelli R, Gott DM, Grob K, Lampi E, Mortensen A, Rivière G, Steffensen I-L, Tlustos C, van Loveren H, Vernis L, Zorn H, Glandorf B, Herman L, Aguilera J, Chesson A (2019) Characterisation of microorganisms used for the production of food enzymes. EFS2 17:e05741. https://doi.org/10.2903/j.efsa.2019.5741

Simpson C (2020) Issues concerning the definition of 'placing on the market' under Regulation (EC) No 178/2002. EFFL 2020:540–547

Turck D, Bresson J-L, Burlingame B, Dean T, Fairweather-Tait S, Heinonen M, Hirsch-Ernst KI, Mangelsdorf I, McArdle H, Naska A, Neuhäuser-Berthold M, Nowicka G, Pentieva K, Sanz Y, Siani A, Sjödin A, Stern M, Tomé D, Vinceti M, Willatts P, Engel K-H, Marchelli R, Pöting A, Poulsen M, Salminen S, Schlatter J, Arcella D, Gelbmann W, de Sesmaisons-Lecarré A, Verhagen H, van Loveren H (2016) Guidance on the preparation and presentation of an application for authorisation of a novel food in the context of Regulation (EU) 2015/2283. EFS2 14. https://doi.org/10.2903/j.efsa.2016.4594

Volpato A (2022) Novel foods in the EU integrated administrative space: an institutional perspective. In: Scaffardi L, Formici G (eds) Novel foods and edible insects in the European Union. Springer International Publishing, Cham, pp 15–36

Zenodo (2023) Updated list of QPS-recommended microorganisms for safety risk assessments carried out by EFSA. https://zenodo.org/record/7554079#.ZGTdKXZByUk. Accessed 16 June 2023

van der Meulen B (2018) The safe food principle: a critical reflection on the key concept of EU food safety law. EIFL Working Paper Series

Chapter 5
Labelling: General Product Naming Requirements

This chapter addresses various (voluntary and obligatory) labelling requirements concerning dairy products obtained from precision fermentation. First, we will describe the existing regulatory framework for naming traditional dairy products. Second, we will provide an overview of their labelling practices and how they differ from dairy products obtained through precision fermentation. Third, we will discuss the legal requirements concerning voluntary labels such as vegan, organic, and clean labels, and examine their applicability to dairy products produced via precision fermentation. Forth, we will present an overview of the usage of health and nutrition claims.

The naming of a foodstuff holds significant importance as it conveys information about the product and shapes consumer expectations. Consequently, product naming plays a crucial role in ensuring that consumers are not misled. The Food Information Regulation (FIR) outlines general requirements for food information and the responsibilities of food business operators. According to Article 7(1) FIR a fair information practice must be followed, ensuring that food information is not misleading. The responsibility to provide accurate product information lies with the operator under whose name or business name the food is marketed or, if that operator is not established in the Union, the importer of that food into the Union market.[1] To prevent misleading information, the FIR establishes both, mandatory and voluntary information for foodstuffs.

Articles 9(1)(a) und 17 FIR stipulate that the name of the food is a mandatory particular. According to Article 17(1) FIR there are three possibilities to determinate the name of the food:

> The name of the food shall be its legal name. In the absence of such a name, the name of the food shall be its customary name, or, if there is no customary name or the customary name is not used, a descriptive name of the food shall be provided.

Legal definitions of these terms can be found in Article 2(2)(n), (o), (p) FIR:

> Article 2(2)(n) FIR: *"'legal name' means the name of a food prescribed in the Union provisions applicable to it or, in the absence of such Union provisions, the name provided for in*

[1] See FIR. Article 8.

F. Ronchetti et al., *The Regulatory Landscape in the EU for Dairy Products Derived from Precision Fermentation*, SpringerBriefs in Law,
https://doi.org/10.1007/978-3-031-49692-9_5

the laws, regulations and administrative provisions applicable in the Member State in which the food is sold to the final consumer or to mass caterers;"

Article 2(2)(o) FIR: "*'customary name' means a name which is accepted as the name of the food by consumers in the Member State in which that food is sold, without that name needing further explanation,"*

Article 2(2)(p) FIR: "*'descriptive name' means a name providing a description of the food, and if necessary of its use, which is sufficiently clear to enable consumers to know its true nature and distinguish it from other products with which it might be confused."*

This report focuses on dairy products, which are subject to legal names.[2] As such, dairy products are required to use these legal names in compliance with the applicable Union provisions in Article 78(2)(c), Annex VII Part. III Reg. (EU) No. 1308/2013 (CMO). According to the given definition in Annex VII Part. III No. 1 CMO,

"Milk" means exclusively the normal mammary secretion obtained from one or more milkings without either addition thereto or extraction therefrom.

However, the term "milk" may be used:

(a) for milk treated without altering its composition or for milk the fat content of which is standardised under Part IV;
(b) in association with a word or words to designate the type, grade, origin and/or intended use of such milk or to describe the physical treatment or the modification in composition to which it has been subjected, provided that the modification is restricted to an addition and/or withdrawal of natural milk constituents.

Additionally, Annex VII Part. III No. 2 CMO holds a definition for milk products:

For the purposes of this Part, "milk products" means products derived exclusively from milk, on the understanding that substances necessary for their manufacture may be added provided that those substances are not used for the purpose of replacing, in whole or in part, any milk constituent.

The following shall be reserved exclusively for milk products.

(a) the following names used at all stages of marketing:

(i) whey,
(ii) cream,
(iii) butter,
(iv) buttermilk,
(v) butteroil,

---

[2] Meisterernst (2019), § 10 marginal no. 133. There is some discussion emerging as to whether the provisions in the CMO conveys the same regulatory objective as Art. 17 FIR. Such a reference of Art. 17 FIR to the provisions of the CMO may hence need not be subject to a suitability test. The Court, however, without further ado references both regimes and connects them to one another, see C-422/16, *TofuTown* [2017] Court of Justice of the European Union ECLI:EU:C:2017:458, marginal no. 43.

(vi) caseins,
(vii) anhydrous milk fat (AMF),
(viii) cheese,
(ix) yogurt,
(x) kephir,
(xi) koumiss,
(xii) viili/fil,
(xiii) smetana,
(xiv) fil;
(xv) rjaženka,
(xvi) rūgušpiens;

(b) names within the meaning of Article 5 of Directive 2000/13/EC or Article 17 of Regulation (EU) No 1169/2011 actually used for milk products.

Annex VII Part III No. 3 CMO holds information about the usage of these terms:

> The term 'milk' and the designations used for milk products may also be used in association with a word or words to designate composite products of which no part takes or is intended to take the place of any milk constituent and of which milk or a milk product is an essential part either in terms of quantity or for characterisation of the product.

Lastly, Annex VII Part III No. 6 CMO states the wide restriction of using these legal names for other products:

> In respect of a product other than those described (…) this Part, no label, commercial document, publicity material or any form of advertising (…) or any form of presentation may be used which claims, implies or suggests that the product is a dairy product. However, in respect of a product which contains milk or milk products, the designation 'milk' or the designations referred to in (…) this Part may be used only to describe the basic raw materials and to list the ingredients in accordance with Directive 2000/13/EC or Regulation (EU) No 1169/2011.

This means that milk is specifically defined as the outcome of mammary secretion, and milk products must be derived from milk. The legal names mentioned in Annex VII Part III No. 2 CMO are exclusively allowed for Milk products. However, these legal names can be combined with other terms to create new designations, such as "Herbal Butter" for butter mixed with different herbs.

If a product does not meet the requirements to be named as milk or a milk product, it is not permitted to use the legal names mentioned above as it could mislead consumers into believing the product contains milk. This is especially relevant for plant-based products.

If milk is used as a raw material for a product, the term milk can be used to describe the product.

The basic name protection for milk products is regulated by the mentioned Union provisions, but Member States arguably have the authority to establish further rules,[3] however, only within the confines set by Union law, especially the FIR. For

[3] Meisterernst (2019), § 15 marginal no. 17.

instance, in Germany there are specific regulations such as the Milk Products Regulation *(Milcherzeugnisverordnung)* and the Cheese Regulation *(Käseverordnung),* which provide definitions and legal names in the sense of Article 2(2)(n) FIR for milk products and cheeses not covered by the GMO,[4] such as "crème fraiche, whipped cream" for milk products and "hard cheese" for cheeses.

Alongside the name of the food, specific details shall accompany the name (see Article 17(5) & Annex VI FIR). These particulars may pertain to the food's physical condition or any specific treatment it has undergone. The inclusion of such information is mandatory if otherwise the omission of such information could mislead the purchaser. Some examples of such treatments, as listed in Annex VI, include powdered, refrozen, freeze-dried, quick-frozen, concentrated, and smoked. However, this list of treatments is not exhaustive.[5] Since the information is required to protect the consumer, it is irrelevant if the accompanying information is disadvantageous.[6]

5.1 Difference to Plant-Based Products

The significance of alternative proteins has steadily increased in recent years, with a particular demand for "plant-based" products. However, there is no legal definition of the term "plant-based" under EU law. For our purposes, plant-based products refer to vegetarian or vegan alternatives designed to replicate the appearance, texture, and taste of animal-based foods.[7]

Presently, most supermarkets offer a diverse range of plant-based products. Initially, when these products were introduced to the market, many producers named them after their conventional animal-based equivalents (e.g., Soymilk, Tofubutter, Cheesesteak) due to their similarity in appearance and taste. However, the judgment "TofuTown"[8] of the Court of Justice of the European Union clarified that under the circumstances of this case such terms should be reserved for "conventional" products and cannot be used for plant-based alternatives.

5.2 The TofuTown Case

In this case, the company "TofuTown" was placing plant-based products on the European market and named them after the conventional product (e.g. *Soyatoo Tofubutter*). The CJEU addressed the issue of whether terms and legal names listed

---

[4] Meisterernst (2019), § 15 marginal no. 18.

[5] Voit and Grube (2016), Article 17 marginal no. 140.

[6] Meisterernst (2022), Article 17 marginal no. 140.

[7] McClements and Grossmann (2022), p. 1.

[8] C-422/16, *TofuTown* [2017] Court of Justice of the European Union ECLI:EU:C:2017:458.

in Article 78(2) and Annex VII CMO (as mentioned above) can be employed for plant-based products. This question arises when the accompanying information clearly indicates that the product is an alternative to the traditional animal-based product. Specifically, terms such as milk, cheese, or yoghurt are under scrutiny in this context.[9]

In its judgment, the CJEU emphasised that the wording used in Annex VII CMO explicitly indicates that *"the term 'milk' cannot, in principle, be lawfully used to designate a purely plant-based product, since milk is, within the meaning of that provision, 'an animal product' (…)".*[10]

The same applies to milk products since they must be made of milk, as it is not allowed to replace the milk in milk products.[11]

The CJEU elucidates its decision by referencing the objectives of these regulations: *"improving the economic conditions for the production and marketing as well as the quality of such products. The application of such standards is therefore in the interest of producers, traders and consumers, to protect consumers and to maintain conditions for allowing competition. Those provisions, in so far as they provide that only the products which comply with the requirements they lay down can be designated by the term 'milk' and the designations reserved exclusively for milk products even if those designations are expanded upon by explanations or descriptions such as those at issue in the main proceedings, contribute to the attainment of those objectives.*[12]

(…) The fact that, in marketing or advertising, the possibility to use the term 'milk' and the designations reserved exclusively for milk products is available only to products which meet the requirements laid down by Annex VII, Part III, to Regulation No 1308/2013 is a guarantee, in particular, to the producers of those products of undistorted conditions for competition, and to consumers of those products, that the products designated by those designations meet all the same standards of quality, both protecting them against any confusion as to the composition of the products they intend to purchase."[13]

However, the CJEU also clarified that there are exceptions for plant-based products to use the legal name. According to Annex VII Part 3 no. 5 sentence 2 it is permissible to use the legal names for milk and milk products: However, *"this provision shall not apply to the designation of products the exact nature of which is*

[9] Questions were stated by a German court in Trier via Preliminary ruling procedure of Article 267 TFEU, see LG Trier, Beschluss vom 28.07.2016 - 7 HK O 20/16.

[10] C-422/16, *TofuTown* [2017] Court of Justice of the European Union ECLI:EU:C:2017:458, marginal no. 23.

[11] C-422/16, *TofuTown* [2017] Court of Justice of the European Union ECLI:EU:C:2017:458, marginal no. 25.

[12] C-422/16, *TofuTown* [2017] Court of Justice of the European Union ECLI:EU:C:2017:458, marginal no. 43.

[13] C-422/16, *TofuTown* [2017] Court of Justice of the European Union ECLI:EU:C:2017:458, marginal no. 48.

clear from traditional usage and/or when the designations are clearly used to describe a characteristic quality of the product."

The European Commission compiled a list of milk and milk products, drawing from indicative lists of products recognised as such by Member States in their respective territory.[14] The CJEU clarifies that all the products featured on this list are permitted to carry a legal name for milk and milk products. For example, coconut milk is included in this list, but almond milk is not. However, soy and tofu are not on the list, and consequently, they are restricted from using the terms "milk" or other legal names for milk products in their product names. Nevertheless, the CJEU emphasises that the Commission's list is subject to potential changes in the future *"in order to satisfy demonstrated needs resulting from evolving consumer demand, technical progress or the need for product innovation (...) However, until today, such an act has not been adopted by the Commission as regards the definitions and designations for milk and milk products."*[15]

As a result of the judgment, plant-based dairy alternatives to animal-based foods are now required to bear more complex names, such as "Soy-Drink" instead of "Soymilk". This type of unintuitive naming adds further complications to the already intricate marketing of these innovative products. Additionally, there is a significant uncertainty surrounding the lawful denomination for animal-free food alternatives to produced through precision fermentation.

5.3 Naming Requirements for Precision Fermentation Products

Regarding precision fermentation products in the dairy sector, the key question revolves around how to name dairy products derived from precision fermentation. During the study, some interviewees (28.58%) had not yet considered naming their products since they were still in the development phase. However, the majority of interview partners had ideas about naming their products. The opinions on naming were varied, with 42.87% agreeing to use the same names as conventional products (e.g., cheese), while 28.58% disagreed.

Arguments in favour of labelling the precision fermented product similarly to the conventional counterpart were based on the fact that cheese derived from precision fermentation has similar components as traditional cheese, and precision fermentation is just another method of producing cheese. Conversely, arguments against identical labelling were that there are still molecular structural differences between conventional cheese and the one obtained through precision fermentation, and using

[14] 2010/791/EU: Commission Decision (2010).

[15] C-422/16, *TofuTown* [2017] Court of Justice of the European Union ECLI:EU:C:2017:458, marginal no. 39.

the same name could be misleading due to the novelty of precision fermentation in the food sector.

The uncertainty surrounding the naming of dairy products obtained from precision fermentation necessitates the clarification of labelling rules. Existing rules for dairy products, as described earlier, are specifically reserved for milk and milk products and clearly defined in the CMO. Precision fermentation more often than not involves genetically engineered microorganisms to create milk products, and although the molecular structure may resemble conventional milk, there is no milking process involved. Establishing a link to the definition of "mammal secretion" through the use of animal genes is tenuous, and it remains unclear how this argumentation would resonate in a judicial dispute.

Conversely, the TofuTown judgment has clearly prohibited the naming of plant-based products imitating conventional dairy products. However, more often than not genetically modified microorganisms are used to obtain precision fermented dairy products, hence they are not produced by plant-based alternatives as conventionally understood not involving food technology. According to the CJEU's wording in TofuTown, the strict prohibition concerns *purely* plant-based products.[16] Products obtained by precision fermentation are not plant-based as understood in the conventional sense. As a consequence, the interpretation offered by the TofuTown judgment is not directly applicable to dairy products.

The judgment highlighted that exceptions from the strict rules for naming dairy products are possible if the product is listed in the Commission's list of exceptions.[17] This list includes for example the term "coconut milk", but not "almond milk". This list has not been updated. At the time of writing it there have been no indications that the Commission is planning to add milk products obtained by precision fermentation to this list of exceptions.

Precision fermentation still needs to find its way between the application of the labelling rules for plant based and conventional foods.

The naming of dairy products is protected by law, and these names cannot be used for other products, regardless of any additional clarification notes or the absence of misleading information.[18]

Due to the strict naming requirements for dairy products, it is not recommended to label precision fermentation products as "cheese" or use any other legal name listed in the CMO. Instead, more neutral names like "slices," "wheel," or "loaf" can be used in connection with a brand name, similar to how they are used for plant-based milk alternatives.

Precision fermentation is a relatively new process in the food sector, and depending on the usage of genetically modified microorganisms, dairy products obtained

---

[16] C-422/16, *TofuTown* [2017] Court of Justice of the European Union ECLI:EU:C:2017:458, marginal no. 23.

[17] 2010/791/EU: Commission Decision (2010).

[18] Meisterernst (2019), § 10 Rn. 132.

through precision fermentation need to be authorised via the GMFR[19] or NFR.[20] Regardless of the authorisation process, products obtained through precision fermentation will have their legal names specified based on the suggestions made by the applicants.

Example labelling for novel foods: When seeking novel food authorisation, the producer is required to submit various mandatory information to EFSA as per Article 10(2) NFR. One such essential piece of information is the name and description of the novel food, as outlined in Article 10(2)(b) of the NFR. Furthermore, the applicant must specify particular labelling requirements that do not mislead consumers or provide a verifiable justification why those elements are not necessary. This requirement is stipulated in Article 10(2)(g) NFR.

This implies that, initially, the responsibility falls upon the producer to determine an appropriate designation for their product. Neither the NFR nor the Commission Implementing Regulation (EU) 2017/2469[21] provide explicit guidance on how the producer should name their novel food product. The EFSA published a Guidance on the preparation and presentation of an application for authorisation of a novel food in the context of Regulation (EU) 2015/2283.[22] In Section 2.2 of this guidance, relevant explanations are provided to assist producers in determining the appropriate information for the naming of their novel food:

> Information on the identity of the novel food should be provided, depending on the class(es) under which the novel food falls. The Panel notes that the proposed classification is based on the chemistry, production process and source of novel foods, for the purpose of the scientific assessment, and is not meant to reflect the regulatory categories outlined in Article 3(2)a of the Regulation. There may be cases where a novel food could be allocated to two or more classes (e.g. 'chemical substances' and 'food produced by a microorganism'). In such cases, the relevant information for all applicable classes should be provided.

During the authorisation procedure, the labelling conditions for a novel food are thoroughly assessed and subsequently defined. It is mandatory for the producer to adhere to these specified labelling requirements if they intend to market their novel food product. Additionally, the Commission holds the authority to include specific labelling requirements aimed at informing the end consumer about particular characteristics or properties of the food. These requirements may pertain to aspects such as the composition, nutritional value, nutritional effects, and intended use of the novel food. The objective is to highlight any differences between the novel food and existing foods, as well as any potential implications for the health of specific population groups. This provision is stipulated in Article 9(3)(b) NFR.

Once the authorisation process is successfully completed, the novel food product will be added to the Union list of novel foods,[23] and a Commission Implementing

---

[19] Regulation (EC) No 1829/2003 (2003).

[20] Regulation (EU) 2015/2283 (2015).

[21] Commission Implementing Regulation (EU) 2017/2469 (2017).

[22] Turck et al. (2016).

[23] Commission Implementing Regulation (EU) 2017/2470 (2017).

Regulation will be issued to officially confirm the authorisation.[24] Both of these documents will outline the specific labelling requirements for the novel food.

This underscores the significance of deciding on an appropriate name for the novel food during its early developmental stages. The Commission considers the chosen name as a basis for their decision-making process. Furthermore, the name of the product plays a crucial role in consumer acceptance, as it can create certain images and preconceptions in the minds of consumers.[25]

Carefully selecting an appropriate name is hence attractive also to food innovators, as producers can positively influence consumer perceptions and foster a favourable reception of their novel food product in the market. Conversely, an ill-suited or misleading name may result in consumer confusion or scepticism. Thus, early consideration of the product name is pivotal for a successful and well-received novel food launch.

Example labelling Genetically Modified Organisms: If the precision fermented product falls within the scope of the GMFR, it is subject to the relative pre-market authorisation procedure as outlined in Article 5 GMFR (as mentioned in the preceding chapter). Similar to the case of novel foods, the application procedure for GM foods entails the preparation of a comprehensive dossier. This dossier must include the specific designation of the food and its detailed specifications, which encompass the transformation event(s) used in the production process.

Additionally, in the scenario where the GMO-food authorisation is processed by the responsible national authority, the opinion paper provided by the authority will also include the designation of the food and its specifications, as mandated by Article 6(5)(b) of the GMFR.

It is important to note that GMO-foods are subjected to specific labelling requirements under Article 13 of the GMFR:

> 1. Without prejudice to the other requirements of Community law concerning the labelling of foodstuffs, foods falling within the scope of this Section shall be subject to the following specific labelling requirements:
>
>   a. where the food consists of more than one ingredient, the words 'genetically modified' or 'produced from genetically modified (name of the ingredient)' shall appear in the list of ingredients provided for in Article 6 of Directive 2000/13/EC in parentheses immediately following the ingredient concerned;
>   b. where the ingredient is designated by the name of a category, the words 'contains genetically modified (name of organism)' or 'contains (name of ingredient) produced from genetically modified (name of organism)' shall appear in the list of ingredients;

[24] As an example, this is the Commission Implementing Regulation on an insect as a novel food: Commission Implementing Regulation (EU) 2021/882 (2021).

[25] Hallman (2020), p. 2268.

c. where there is no list of ingredients, the words 'genetically modified' or 'produced from genetically modified (name of organism)' shall appear clearly on the labelling;
d. the indications referred to in (a) and (b) may appear in a footnote to the list of ingredients. In this case they shall be printed in a font of at least the same size as the list of ingredients. Where there is no list of ingredients, they shall appear clearly on the labelling;
e. where the food is offered for sale to the final consumer as non-pre-packaged food, or as pre-packaged food in small containers of which the largest surface has an area of less than 10 cm^2, the information required under this paragraph must be permanently and visibly displayed either on the food display or immediately next to it, or on the packaging material, in a font sufficiently large for it to be easily identified and read.

2. In addition to the labelling requirements referred to in paragraph 1, the labelling shall also mention any characteristic or property, as specified in the authorisation, in the following cases:

   a. where a food is different from its conventional counterpart as regards the following characteristics or properties:

      (i) composition;
      (ii) nutritional value or nutritional effects;
      (iii) intended use of the food;
      (iv) implications for the health of certain sections of the population;

   b. where a food may give rise to ethical or religious concerns.

3. In addition to the labelling requirements referred to in paragraph 1 and as specified in the authorisation, the labelling of foods falling within the scope of this Section which do not have a conventional counterpart shall contain appropriate information about the nature and the characteristics of the foods concerned.

According to Article 13 GMFR, ingredients that contain or consist of GMOs or are produced from or contain ingredients produced from GMOs must be labelled clearly *('genetically modified' or 'produced from genetically modified (name of the ingredient)').*

Similar to the process of novel food authorisation, the producer must take utmost care in selecting an appropriate name and creating accurate labelling for the genetically modified product, as these aspects significantly influence the approval decision. Once the authorisation process is successfully completed, the GM foods will be registered in the EU Register of authorised GMOs.[26] This register serves as a comprehensive database that documents and lists all the genetically modified products that have been approved for commercialisation within the European Union. By

[26] European Commission (2023a).

being included in this register, the authorised GMOs gain recognition and are officially authorised for market distribution within the EU's regulatory framework.

5.4 Vegan

The following labelling question pertains to whether products derived from precision fermentation can be labelled as "vegan." As discussed earlier in the context of naming issues, it is important to note that products obtained through precision fermentation are not purely plant-based, as no plant products are involved in the fermentation process. Instead, the cells utilised in this fermentation process are derived from microbial metabolism and are bioidentical to those produced by mammals.

Due to the absence of an established legal definition for terms like "vegan" or "vegetarian" in EU law,[27] there is a degree of flexibility in how these labels can be used. As a result, it remains possible for products derived from precision fermentation to be labelled as "vegan."[28] To address the legal gap caused by the absence of a standardised definition of "vegan" or "vegetarian" food in EU law, certain Member States have taken the initiative to establish their own definitions. In Germany, for example, in 2018 the Ministry of Food and Agriculture published *Guidelines for vegan and vegetarian foods similar to foods of animal origin*.[29] However, these guidelines are not legally binding.[30] The lack of a standardised definition of "vegan" or "vegetarian" food in EU law creates uncertainties for producers of plant-based alternatives, particularly in the precision fermentation sector. This uncertainty arises from the fact that, in principle, precision fermentation involves the use of cells that are bioidentical to animal-derived cells to obtain the final product. Furthermore, independent private bodies offer certifications for vegan and vegetarian products.[31] These certifications provide an alternative means for producers to label their products as vegan or vegetarian based on specific criteria set forth by these organizations. These criteria had been developed at a "Verbraucherschutzministerkonferenz" (conference of consumer protection ministers) in 2016.[32]

The definition proposed by this conference[33] stipulates, that

> Food is vegan, when they are not products of animal origin and in which at all stages of production and processing no

---

[27] So far, the Commission did not utilise its mandate to adopt implementing acts concerning information related to suitability of a food for vegetarians or vegans pursuant to Article 36 (3)(b) FIR.

[28] Leialohilani and de Boer (2020), p. 265.

[29] Bundesministerium für Ernährung und Landwirtschaft (2018).

[30] Meisterernst (2021), marginal no. 1.

[31] E.g. The European Vegetarian Union (V-Label), see European Vegetarian Union (2021).

[32] For the German Guideline Definition see: Bundesministerium für Ernährung und Landwirtschaft (2018), Sec. 1.1.1.; For the V-Label see European Vegetarian Union (2023).

[33] Verbraucherschutzministerkonferenz (2016).

(1) ingredients (including additives, carriers, flavourings and enzymes) or
(2) processing aids or
(3) non-food additives that are used in the same way and for the same purpose as processing aids,

of animal origin are added or used in processed or unprocessed form.

Microorganisms (bacteria, yeasts and fungi) are of non-animal origin and may also be used in foods that are advertised as "vegan".

The definition established at this conference offers a significant advantage in that it takes a process-oriented approach to understanding vegan food. This means that the definition acknowledges that, in some cases, there may be unavoidable and unintended contamination with animal components due to the limitations of current technology and manufacturing processes.[34]

The European Vegetarian Union (EVU) exclusively awards the V-Label certificate to products that align with the definition established in the "Verbraucherschutzministerkonferenz" conference. This means that only products that meet the process-oriented understanding of vegan food, including the consideration of unavoidability in certain cases of contamination, are eligible for the V-Label certification.

However, obtaining the V-Label certification involves meeting additional criteria set by the EVU. For example, products that are required to be labelled as genetically modified organisms (GMOs) are not eligible for the V-Label certification. Therefore, if a product has been authorised as a GMO-Food, it is officially unable to receive the V-Label certification, as it does not meet this specific criterion.[35]

The EVU exclusively awards the V-Label certificate to products that align with the definition established in the "Verbraucherschutzministerkonferenz" conference. This means that only products that meet the process-oriented understanding of vegan food, including the consideration of unavoidability in certain cases of contamination, are eligible for the V-Label certification.

However, obtaining the V-Label certification involves meeting additional criteria set by the EVU. For example, products that are required to be labelled as genetically modified organisms (GMOs) are not eligible for the V-Label certification. Therefore, if a product has been authorised as a GMO-Food, it is officially unable to receive the V-Label certification, as it does not meet this specific criterion.

Nonetheless, even if the product is ineligible for the V-Label certification, it can still be advertised as "vegan" as long as it fulfils the process-oriented definition established by the "Verbraucherschutzministerkonferenz" conference. The term

[34] See also: European Vegetarian Union (2021) *"Products carrying this information meet the criteria of the V-Label provided that no animal ingredients are deliberately added at any point during production. (…) According to the criteria of the V-Label, total contamination with all verifiable animal ingredients in the end product must not exceed 0.1% (1 g/kg). Other rules based on a "technically feasible detection limit" may apply for certain product groups."*; Jäger (2019), p. 251; More Information (in German): proveg international (2018).

[35] European Vegetarian Union (2023).

"vegan" itself is not a protected claim, meaning that producers can use it to describe their products if they meet the criteria of the definition.

Contrary to plant-based alternatives, products obtained through precision fermentation utilise cells produced in the laboratory to create food. This raises the question of whether these products can still be labelled as "vegan." To gain insights, interviews were conducted with experts, and their responses were quite similar: All experts indicated that it is indeed possible to label products obtained through precision fermentation as vegan. Their reasoning was centred around the fact that precision fermentation offers a more sustainable method of food production compared to conventional methods, as it does not involve livestock farming and requires only a minimal number of cells. While the majority of experts supported the possibility of labelling precision-fermented products as vegan, some acknowledged that there may be pushback from individuals who define veganism more strictly and reject precision fermentation based on its use of bioidentical cells. To address potential concerns and provide additional transparency to consumers, some interviewees suggested using additional claims such as "animal-free" or "slaughtering-free" to inform customers that no animals are involved in the production process. It is worth noting that the V-Label certification, which is recognized for vegan products, allows for a small contamination of animal ingredients up to 0.1% (1 g/kg).[36] Given that products obtained through precision fermentation use cells that are bioidentical to animal-derived cells, it is likely that these products would still meet the strict criteria of the V-Label certification and be eligible for the label.

During the interviews, one of the experts emphasised that the core objectives of the V-Label certification are centred around sustainability and animal welfare. Considering that milk products obtained through precision fermentation align with these specific requirements, it is recommended to contact the V-Label team and seek their advice.

5.5 Organic Certification

The question at hand is whether products obtained from precision fermentation can obtain the EU organic certification and be labelled with the "organic" label. The organic certification process is governed by Regulation (EU) No 848/2018 on organic production and labelling of organic products.[37] This Regulation outlines the principles of organic production and establishes rules pertaining to organic production, certification, and the use of indications related to organic production in labelling and advertising. In light of this, it is essential for food business operator using precision fermentation to carefully assess whether their production process meets the specific requirements laid out in Regulation (EU) No 848/2018. Compliance

[36] European Vegetarian Union (2023); Jäger (2019), p. 251; proveg international (2018).

[37] Regulation (EU) 2018/848 (2018).

with these regulations will determine whether their products are eligible for the European organic certification and can bear the "organic" label on their packaging.[38]

The objectives of this regulation are stated in Article 4:

(a) contributing to protection of the environment and the climate;
(b) maintaining the long-term fertility of soils;
(c) contributing to a high level of biodiversity;
(d) substantially contributing to a non-toxic environment;
(e) contributing to high animal welfare standards and, in particular, to meeting the species-specific behavioural needs of animals;
(f) encouraging short distribution channels and local production in the various areas of the Union;
(g) encouraging the preservation of rare and native breeds in danger of extinction;
(h) contributing to the development of the supply of plant genetic material adapted to the specific needs and objectives of organic agriculture;
(i) contributing to a high level of biodiversity, in particular by using diverse plant genetic material, such as organic heterogeneous material and organic varieties suitable for organic production;
(j) fostering the development of organic plant breeding activities in order to contribute to favourable economic perspectives of the organic sector.

The concept of organic farming is intended to promote sustainable food production. The European Green Deal[39] and the Farm-to-Fork Strategy[40] both prioritised sustainable (food) production as one of the main objectives of EU policy. The Farm-to-Fork strategy specifically emphasises the importance of increasing organic farming as a means to achieve these objectives.[41] However, the current regulatory framework for organic production is not very innovation-friendly. For instance, certain innovative approaches such as vertical farming[42] which offer potential solutions to counter the negative impacts of traditional agriculture, are not allowed to be certified as organic because organic production must take place in soil.[43]

The anti-innovation approach to organic food production, where certain innovative processes, such as precision fermentation, are not allowed to be labelled as organic, presents significant challenges. Many innovative production processes and techniques aim to produce more sustainable and environmentally friendly products, which aligns with the core principles of organic farming.

---

[38] See Article 33(1) Regulation (EU) 2018/848 (2018).

[39] European Commission (2019).

[40] European Commission (2020).

[41] European Commission (2020), p. 3.

[42] Vertical farming means, that agricultural products will be farmed upwards rather then outwards to increase the crop per unit area of land, see: Beacham et al. (2019).

[43] Neslen (2021).

During the interviews conducted, all experts expressed a unanimous opinion that dairy products obtained through precision fermentation cannot be labelled as organic. One interviewee cited the example of vertical farming, indicating that since precision fermentation is not "conventionally harvested" like traditional organic agriculture, it falls outside the scope of organic certification. Another interviewee pointed out the "GMO-history of the product," referring to the use of bioidentical cells, which is considered a GMO process. This expert argued that the GMO aspect makes it impossible to label these products as organic.

If a fermented product falls under the scope of GMO regulations it cannot be simultaneously labelled as organic.[44] According to Article 11(1) Regulation (EU) No 848/2018 states that "products produced from GMOs and produced by GMOs shall not be used in organic food." However, there is an important distinction to be made between "produced from GMOs" and "produced with GMOs". Organic production of foods using genetically modified microorganisms (GMMs) could potentially become a common practice if the food is considered to be produced "with" GMOs.[45] This nuance may open up possibilities for further exploration of precision fermentation and other innovative methods within organic farming, while still adhering to the principles of sustainability and environmental protection.

Under the current status of organic law, products derived from precision fermentation face challenges in obtaining the EU Organic Label due to the novelty of the process and the absence of detailed production rules for such innovative methods in the Commission's list for detailed production rules.[46] Under the current status of organic law, products derived from precision fermentation would not be allowed to be labelled as organic with the EU Organic Label. However, it's important to note that based on the EU Organic Certification, several Member States developed their own national labels for organic certification, such as the German "Bio-Siegel".[47] Additionally, there are several private labels for organic certification, gold-plating the EU Organic Label, such as Demeter[48] or Bioland.[49]

5.6 Clean Labels

In recent years, the use of "clean labels" on food products has become more prevalent as they can significantly influence consumer food choices.[50] Clean labels can encompass various types, either highlighting the absence of certain ingredients like

---

[44] See Regulation (EU) No 2018/848. Articles 11, 30(4) and Recital 23.

[45] Purnhagen et al. (2021).

[46] See Regulation (EU) No 2018/848. Articles 21, 54 and Annex II.

[47] Bundesministerium für Ernährung und Landwirtschaft (2022).

[48] Demeter e.V. (2023).

[49] Bioland e.V. (2023).

[50] Cao and Miao (2023), p. 13.

additives or flavourings,[51] or promoting the use of only a few simple, recognisable ingredients and traditional production processes.[52] Examples of clean labels include claims such as "GMO-free" or "Eco-friendly". During the interviews, it was observed that many participants expressed interest in using clean labels on their products. Specifically, they were keen on advertising their products as "animal-free" or "free from animals," similar to claims commonly used in the cosmetic industry, such as "cruelty-free" or "free from animal testing." However, it is important to note that at the EU level, there are no specific regulations governing these claims for food products.[53] Despite the absence of specific Union standards, clean labels must adhere to the requirements regarding voluntary labelling of the FIR, such as to ensure that they are not misleading to consumers. Even in the absence of specific Union standards, all clean labels must fulfil the requirements of the FIR and shall therefore not be misleading. As the precision fermentation process involves neither animals nor products directly obtained from animals using clean labels like "animal-free" on these foodstuffs does not appear to be misleading.

5.7 Claims

Next to vegan and clean labels, food products can also bear various claims that are regulated by EU law. In this section, we will provide essential information regarding the use of health and nutrition claims and geographical indications.

5.7.1 Health and Nutrition Claims

Health and nutrition claims are voluntary information which food business operator can choose to provide. These claims are governed by Regulation (EC) No 1924/2006 on nutrition and health claims (HNCR).[54] A claim is defined as "any message or representation, which is not mandatory under Community or national legislation, including pictorial, graphic or symbolic representation, in any form, which states, suggests or implies that a food has particular characteristics".[55]

The claims that are made must, inter alia, not "be false, ambiguous or misleading."[56]

[51] Cao and Miao (2023), p. 13; Lebensmittelverband Deutschland (2023).

[52] Molitorisová et al. (2021), p. 76.

[53] Some private organisations like PETA do have lists of companies that do not test their products on animals and do label their cosmetic products as e.g. cruelty-free or free from animal testing, see BeautyWithoutBunnies (2023).

[54] Regulation (EC) No 1924/2006 (2006).

[55] Regulation (EC) No 1924/2006. Article 2(2) No. 1.

[56] Regulation (EC) No 1924/2006. Article 3.

The provisions of the NHCR divides claims into two groups: health claims and nutrition claims.

Nutrition claims are: "any claims which state, suggest or imply that a food has particular beneficial nutritional properties" due to features listed in the NHCR,[57] such as "high fiber".

Health claims are: "any claims that state, suggest or imply that a relationship exists between a food category, a food or one of its constituents and health,"[58] such as Calcium is needed for the maintenance of normal bones".

A list of all authorised nutrition claims can be found in the Annex of the NHCR, as specified in its Article 8. Food business operators can use these claims if their products meet the conditions outlined in the Annex.[59] Unlike health claims, there is no separate authorisation procedure required to use nutrition claims. If a product aligns with the listed claims, it can be employed accordingly.[60] Nutrition claims can also be utilised to compare foods with the same category, considering a range of food products within the same category (see Article 9). Some interviewees expressed their interest in not merely imitating conventional products with precision-fermented alternatives, but also in enhancing the nutritional value of their products and potentially using comparative claims. As precision-fermented dairy products fall within the same category as their conventional counterparts, it should be permissible to compare their nutritional values.

Health claims must undergo an authorisation process before they can be placed on the product.[61] Food business operators can use authorised health claims if the conditions specified for those claims are applicable to their products. Conversely, non-authorised health claims are not allowed to be used. It is crucial to understand that health claims pertain not to the product itself, but rather to the specific nutrient, substance, food, or food category for which they have been authorised. For example, it is not permissible to claim that cheese plays a role in the process of cell division and specialisation. Instead, the authorised claim is: "Calcium has a role in the process of cell division and specialisation." The health relationship is between calcium and the regulation of cell division and differentiation.[62]

---

[57] Regulation (EC) No 1924/2006. Article 2(2) No. 4.

[58] Regulation (EC) No 1924/2006. Article 2(2) No. 5.

[59] For example, the claim "High Protein" can be used for products where at least 20% of the energy value of the food is provided by protein.

[60] See Article 15 NHCR for the difference to health claims.

[61] There is a difference between specific and non-specific health claims: Specific Health claims are subject to approval (See Regulation (EC) No 1924/2006. Article 10(1)) and are prohibited until authorisation and are then included in a list of claims. Characteristic for specific health claims is a scientific proof of the interdependency between the substance and body function. Non-specific health claims (e.g.: Good for health) are because of its general formulation no subject of approval (See Reg. (EC) No. 1924/2006. Article 10(3)) They therefore require coupling with authorised specific health claims instead of authorisation. For more information see: Holle and Hüttebräuker (2018) Article 10 marginal no. 12 f.

[62] All authorised and non-authorised health claims and their health relationship can be seen in the EU register of health claims, see European Commission (2023c).

As mentioned before, some interviewees were also interested in using health claims if applicable to their precision-fermented products. To do so, they would need to ensure that the claims are authorised and meet the necessary conditions set forth in the regulation.

In conclusion, precision-fermented food producers can use authorised nutrition claims without a separate authorisation procedure if their products meet the specified conditions. For health claims, a prior authorisation is required, and only authorised claims can be used, referring specifically to the nutrient or substance for which they have been approved. Ensuring compliance with these regulations will enable food business operators to effectively use both nutrition and health claims in marketing their precision-fermented products, providing valuable information to consumers about their nutritional qualities and potential health benefits.

5.7.2 Designation of Origin

The EU has implemented a scheme for protected designations of origin and (PDO) and protected geographical indications (PGI) to provide support to producers of products associated with specific geographical areas.[63] The rules concerning the use of these specific labels are outlined in Regulation (EU) No 1151/2012 on quality schemes for agricultural products and foodstuffs.[64] According to Article 5 Regulation (EU) No 1151/2012, PDO and PGI are two distinct categories with specific criteria for their use:

A PDO can be used for products that

(1) originate in a specific place, region or, in exceptional cases, country;
(2) whose quality or characteristics are essentially or exclusively due to a particular geographical environment with its inherent natural and human factors; and
(3) the production steps of which all take place in the defined geographical area.[65]

In contrast to that, a PGI is a product

(1) originating in a specific place, region or country;
(2) whose given quality, reputation or other characteristic is essentially attributable to its geographical origin; and
(3) at least one of the production steps of which takes place in the defined geographical area.[66]

---

[63] See Regulation (EU) No 1151/2012. Article 4, these objectives should be fulfilled by e.g. securing fair returns for the qualities of their products, ensuring uniform protection of the names as an intellectual property right in the territory of the Union and providing clear information on the value-adding attributes of the product to consumers.

[64] Regulation (EU) No 1151/2012 (2012).

[65] See Regulation (EU) No 1151/2012. Article 5(1).

[66] See Regulation (EU) No 1151/2012. Article 5(2).

To qualify for a PDO, the entire production process must occur within the designated geographical area. Conversely, for a PGI, it is sufficient for at least one step of the production process to take place in that defined area.

If a name is registered, the product is protected against:

(1) any direct or indirect commercial use of a registered name in respect of products not covered by the registration, where those products are comparable to the products registered under that name or where using the name exploits, weakens or dilutes the reputation of the protected name, including when those products are used as an ingredient;
(2) any misuse, imitation or evocation, even if the true origin of the products or services is indicated or if the protected name is translated or accompanied by an expression such as 'style', 'type', 'method', 'as produced in', 'imitation' or similar, including when those products are used as an ingredient;
(3) any other false or misleading indication as to the provenance, origin, nature or essential qualities of the product that is used on the inner or outer packaging, advertising material or documents relating to the product concerned, and the packing of the product in a container liable to convey a false impression as to its origin;
(4) any other practice liable to mislead the consumer as to the true origin of the product.[67]

All products eligible to bear a PDO or PGI designation are listed in the EU's geographical indications register known as eAmbrosia.[68]

The EU's scheme for protected designations of origin and protected geographical indications aims to recognise and support products with strong ties to specific geographical regions. Producers seeking these labels must meet the defined criteria, and understanding the distinctions between PDO and PGI is crucial in determining the appropriate label for their products. By adhering to the regulation, producers can benefit from the protected status, which promotes the unique qualities and reputation of their products linked to a particular geographic origin.

Most of our interviewees are involved in creating foodstuffs that replicate conventional dairy products, particularly cheese. Upon consulting eAmbrosia, it becomes evident that there are 255 registered cheese products, such as Gruyère, recognized as a French PGI, or Allgäuer Emmentaler, acknowledged as a German PDO. However, it is essential to understand that not all cheeses are alike, and precision fermentation may be employed to imitate various cheese types. Consequently, several interviewees expressed interest in potentially seeking PDOs or PGIs for their products in the future.

One argument put forth was the concept of maintaining local production, suggesting that bioreactors used in the fermentation process could be situated within the specific region, aligning with the requirements of PDOs or PGIs. While utilising

67 See Regulation (EU) No 1151/2012. Article 13(1).

68 European Commission (2023b).

these designations could contribute to enhancing the acceptance of precision-fermented products, at present, the use of quality schemes for precision-fermented dairy alternatives is not feasible. This is because the specifications tied to PDOs and PGIs are stringent and primarily centred around location- and tradition-based production processes.

As a result, it remains highly questionable whether the EU would permit the use of PDOs or PGIs for precision-fermented dairy products in the future. The complexity and specificity of the existing schemes may not align adequately with the innovative and cutting-edge nature of precision fermentation processes, creating challenges for their application. Until more flexible and inclusive regulations are introduced, it seems unlikely that PDOs or PGIs will be extended to precision-fermented dairy products, leaving producers with alternative strategies to gain recognition and acceptance for their innovative offerings.

References

Beacham AM, Vickers LH, Monaghan JM (2019) Vertical farming: a summary of approaches to growing skywards. J Hortic Sci Biotechnol 94:277–283. https://doi.org/10.1080/14620316.2019.1574214

BeautyWithoutBunnies (2023) Can I Trust a Product Label That Says, 'No Animal Testing'?

Bioland e.V. (2023) Wir sind die Treibende Kraft für die Landwirtschaft der Zukunft. https://www.bioland.de/verbraucher. Accessed 17 June 2023

Bundesministerium für Ernährung und Landwirtschaft (2018) Leitsätze für vegane und vegetarische Lebensmittel mit Ähnlichkeit zu Lebensmitteln tierischen Ursprungs: Neufassung vom 04. Dezember 2018 (BAnz AT 20.12.2018 B1, GMBl 2018 S. 1174). https://www.bmel.de/SharedDocs/Downloads/DE/_Ernaehrung/Lebensmittel-Kennzeichnung/LeitsaetzevegetrarischeveganeLebensmittel.pdf?__blob=publicationFile&v=5. Accessed 9 June 2023

Bundesministerium für Ernährung und Landwirtschaft (2022) Bio-Siegel. https://www.bmel.de/DE/themen/landwirtschaft/oekologischer-landbau/bio-siegel.html. Accessed 17 June 2023

Cao Y, Miao L (2023) Consumer perception of clean food labels. BFJ 125:433–448. https://doi.org/10.1108/BFJ-03-2021-0246

Demeter e.V. (2023) Konsequent Bio. https://www.demeter.de/. Accessed 17 June 2023

European Commission (2019) The European Green Deal. https://eur-lex.europa.eu/legal-content/EN/TXT/?uri=celex%3A52019DC0640. Accessed 19 January 2024

European Commission (2020) A Farm to Fork Strategy for a fair, healthy and environmentally-friendly food system. https://eur-lex.europa.eu/legal-content/EN/TXT/?uri=CELEX%3A52020DC0381. Accessed 19 January 2024

European Commission (2023a) eAmbrosia: the EU geographical indications register. https://ec.europa.eu/info/food-farming-fisheries/food-safety-and-quality/certification/quality-labels/geographical-indications-register/. Accessed 9 June 2023

European Commission (2023b) EU register of health claims. https://ec.europa.eu/food/food-feed-portal/screen/health-claims/eu-register. Accessed 9 June 2023

European Commission (2023c) Genetically modified organisms. https://webgate.ec.europa.eu/dyna2/gm-register/. Accessed 9 June 2023

European Vegetarian Union (2021) V-Label. https://www.euroveg.eu/v-label/. Accessed 9 June 2023

European Vegetarian Union (2023) Häufige Fragen. https://www.v-label.eu/de/faq. Accessed 9 June 2023

Hallman WK (2020) An empirical assessment of common or usual names to label cell-based seafood products. J Food Sci 85:2267–2277. https://doi.org/10.1111/1750-3841.15351
Holle M, Hüttebräuker A (eds) (2018) HCVO: Health-Claims-Verordnung: Verordnung (EG) Nr. 1924/2006 über nährwert- und gesundheitsbezogene Angaben über Lebensmittel
Jäger V (2019) Kritische Betrachtung der Vegan-/Vegetarisch-Leitsätze aus Sicht der Hersteller. LMuR:249–255
Lebensmittelverband Deutschland (2023) Clean Lables. https://www.lebensmittelverband.de/de/lebensmittel/werbung/clean-labels. Accessed 9 June 2023
Leialohilani A, de Boer A (2020) EU food legislation impacts innovation in the area of plant-based dairy alternatives. Trends Food Sci Technol 104:262–267. https://doi.org/10.1016/j.tifs.2020.07.021
McClements DJ, Grossmann L (2022) Next-generation plant-based foods: design, production, and properties, 1st edn. Springer International Publishing; Imprint Springer, Cham
Meisterernst A (2019) Lebensmittelrecht. Studium und Praxis. C.H. Beck, München
Meisterernst A (2021) Lebensmittel- und Futtermittelgesetzbuch § 15 Deutsches Lebensmittelbuch. In: Streinz R, Meisterernst A (eds) BasisVO / LFGB: Verordnung (EG) Nr. 178/2002 Lebensmittel- und Futtermittelgesetzbuch
Meisterernst A (2022) 113. Verordnung (EU) Nr. 1169/2011 des Europäischen Parlaments und des Rates betreffend die Information der Verbraucher über Lebensmittel, LMIV. In: Meisterernst A, Sosnitza O (eds) Lebensmittelrecht: Werkstand: 185. EL Dezember 2022, vol 185
Molitorisová A, Monaco A, Purnhagen KP (2021) An analysis of the regulatory framework applicable to products obtained from mushroom and mycelium. SSRN J. https://doi.org/10.2139/ssrn.3955899
Neslen A (2021) Vertical farming's sky-high ambitions cut short by EU organic rules: startups argue tough rules are blocking innovation in a sector that shares Brussels' green goals. https://www.politico.eu/article/vertical-farming-eu-organic-rules-startups/. Accessed 9 June 2023
proveg international (2018) "Kann Spuren von Milch und EI Enthalten" – Trotzdem Vegan? https://proveg.com/de/blog/kann-spuren-von-milch-und-ei-enthalten-trotzdem-vegan/. Accessed 9 June 2023
Purnhagen KP, Clemens S, Eriksson D, Fresco LO, Tosun J, Qaim M, Visser RGF, Weber APM, Wesseler JHH, Zilberman D (2021) Europe's farm to fork strategy and its commitment to biotechnology and organic farming: conflicting or complementary goals? Trends Plant Sci 26:600–606. https://doi.org/10.1016/j.tplants.2021.03.012
Turck D, Bresson J-L, Burlingame B, Dean T, Fairweather-Tait S, Heinonen M, Hirsch-Ernst KI, Mangelsdorf I, McArdle H, Naska A, Neuhäuser-Berthold M, Nowicka G, Pentieva K, Sanz Y, Siani A, Sjödin A, Stern M, Tomé D, Vinceti M, Willatts P, Engel K-H, Marchelli R, Pöting A, Poulsen M, Salminen S, Schlatter J, Arcella D, Gelbmann W, de Sesmaisons-Lecarré A, Verhagen H, van Loveren H (2016) Guidance on the preparation and presentation of an application for authorisation of a novel food in the context of Regulation (EU) 2015/2283. EFS2 14. https://doi.org/10.2903/j.efsa.2016.4594
Verbraucherschutzministerkonferenz (2016) Definitionen vegan-vegetarisch. https://www.verbraucherschutzministerkonferenz.de/documents/top20_definition_vegan_und_vegetarisch_1510317864.pdf. Accessed 9 June 2023
Voit W, Grube M (2016) Lebensmittelinformationsverordnung: Kommentar [LMIV; VO (EU) Nr. 1169/11, 2nd edn. Beck-Online Bücher. C.H. Beck, München

Chapter 6
Comparative Analysis

In this chapter we undertake two concise comparative analyses. First, we examine whether the findings mentioned earlier can be extrapolated to other products, investigating if the legal framework for placing precision fermented dairy alternatives on the market is also applicable to other foods derived from or obtained using this technology. Secondly, we will compare the EU legal requirements and authorisation framework for precision-fermented foods with those currently established in Singapore and the U.S. By conducting this comparison, we aim to gain insights into how different jurisdictions approach the regulation and authorisation of precision-fermented products. Understanding these variations can shed light on the potential challenges and opportunities for market entry and acceptance in different regions, and provide valuable information for food producers considering international expansion.

6.1 Comparisons with Other PF-Products

Precision fermentation extends beyond the production of dairy alternatives. In the food industry, this technology serves various purposes and objectives. Some of its applications include enhancing the taste of plant proteins, developing microbial proteins with improved nutritional properties, and, as demonstrated earlier with casein, generating non-animal proteins that are bioidentical to their animal counterparts.[1] Precision fermentation's utilisation is expanding across various food categories, including egg and meat alternatives.[2]

Would other food products obtained through the same technology be subject to similar regulations as the dairy alternatives discussed in this report? Our focus is on pre-market authorisation and denomination requirements only. It is crucial to note that all other obligations for introducing food products to the market under EU law, and as applicable, national law (such as safety, hygiene, labelling, and all other requirements), must be equally adhered to for all food types.

[1] BRAIN Biotech AG (2023).

[2] Eder (2023); Schweizer (2022); The EVERY Company (2023).

F. Ronchetti et al., *The Regulatory Landscape in the EU for Dairy Products Derived from Precision Fermentation*, SpringerBriefs in Law,
https://doi.org/10.1007/978-3-031-49692-9_6

When initiating **pre-market authorisation** procedures, food business operators face the primary challenge of classifying the food product. For all innovative foods that incorporate precision fermentation in their production, classification must be carried out following the procedure described in Chap. 2. The pre-market authorisation process hinges on the detection of rDNA from a GMM in the food.

- If rDNA of the employed GMM is detected in the final product (also if that classifies as ingredient), the food is considered a GM food and falls under the scope of the GMOR as "food produced from a GMO".[3]
- If no rDNA of the GMM is detected in the final product, the food is deemed to only be produced "with" a GMO[4] and falls under the scope of the NFR. The same applies to precision fermented foods for which a non-genetically engineered microorganism is used.

It is therefore impossible to give general advice on the applicable legal regime based on the type of food, as the classification into GM or novel food must be performed on a case-by-case basis.

Regarding all other types of precision fermented foods, their **denomination** is governed by the Food Information Regulation (FIR). The name of the food is considered one of the mandatory pieces of information, and it must not be misleading.[5] However, determining whether a name is misleading or not can be challenging and subjective. Ideally, food business operators (FBOs) should find a balance between informing consumers about the product's substitute nature while avoiding any misinterpretation of its actual content or origin. This could be achieved by refraining from using evocative depictions or larger font sizes when indicating the conventional food they are imitating.

The requirement to use the legal name, or in absence thereof a customary name, or a descriptive name if neither is available, is applicable to all foods. When considering whether a precision fermented alternative can be named similarly to its conventional counterpart, it is crucial to verify the presence of relevant legal names and their respective definitions. As discussed earlier, dairy alternatives cannot be labelled with designations that are legally reserved for dairy products according to the CMO.

Requirements for artificial meat products or plant-based meat substitutes differ slightly. While certain specific names referring to the meat's origin, such as "veal," are reserved,[6] names related to the presentation of meat, such as "burger," "sausage," and "steak," are not legally defined, and there have been no recent CJEU judgment regarding these nomenclatures. On 23 October 2020, the European Parliament voted against extending strict protection (in the form of a ban) for meat-related names in non-animal-based alternatives under the CMO.[7] Although there

[3] van Goethem (2016).

[4] van Goethem (2016).

[5] FIR. Article 9 (1)(a).

[6] FIR. Annex VII Part B (17); CMO. Annex VII Part I.

[7] European Parliament (2020); Pig Progress (2021).

may be a general prohibition on misleading practices, this obstacle could be overcome by providing distinct and clear descriptive labelling.[8] However, food business operators interested in using "meaty" names for their innovative products must verify the absence of relevant provisions in Member States' law or similar national requirements. For instance, Germany has developed "Guiding principles for vegan and vegetarian foods similar to foods of animal origin",[9] and in France Decree no. 2022-947 of 29 June 2022 initially established a ban on using terminology specific to meat and fish sectors (with the exception of "burger") for products not belonging to the animal kingdom. However, this ban was suspended by the Consil d'État on 27 July 2022.[10]

Similar conditions apply to egg alternatives. The CMO includes the term "eggs"[11] as one of the definitions or designations laid down in Annex VII which "may be used in the Union only for the marketing of a product which conforms to the corresponding requirements laid down in that Annex."[12] However, Annex VII Part VI refers to "eggs of hens of the *Gallus gallus* species" only, and not to "eggs" in general. The CMO provides further definitions concerning the egg sector, which solely include more specific designations, such as "eggs in shell" or "eggs for hatching".[13] More stringent definitions of "eggs", which tie the term to specific animals, exist within the meaning of other EU regulations, but are restricted to the corresponding scope.[14] Plant-based egg alternatives bearing denominations such as "vegan eggs" (e.g. "das vegane Ei"), "vEGGs" or "organic plant-based egg substitute" ("Substituto vegetal de huevo ecologico") are already widely spread on the EU market. As of now, these products have not been challenged before the CJEU. Again, the avoidance of consumer misleading is crucial[15] and must be considered when deciding upon the denomination of the product. Additional information (like "vegan") or the use of the English term instead of a translation in the official language of the MS the product is sold may therefore be helpful. Potential requirements at MS level should be taken into account.

---

[8] Sochirca (2018).

[9] Bundesministerium für Ernährung und Landwirtschaft; retrieved: Bundesministerium für Ernährung und Landwirtschaft (2018).

[10] Southey (2022).

[11] CMO. Article 78(1)(e).

[12] CMO. Article 78(2).

[13] CMO. Annex II Part VII.

[14] Examples of this stringent production requirement are "*produced by hens of the species Gallus gallus*" in Article 1 (k) of Regulation (EC) No 589/2008 (2008).

[15] FIR. Article 7.

6.2 Comparisons with Other Countries

When we asked our interviewees about their preferred regulatory framework to work in, they often mentioned the USA and Singapore as more open and innovative-driven regarding novelties in the food sector. Therefore, in this section, we will provide a broad overview of how the legal frameworks in Singapore and the USA differ from the EU framework when it comes to placing precision fermentation products on the market.

Singapore is a City-State with a size of 719 qm^2.[16] This is approximately the same size as Hamburg.[17] Singapore is profiting from maritime trading routes, and therefore, more than 90% of the food consumed in Singapore is imported.[18] By 2014, only 1% of the Singaporean land was used for farming.[19] To ensure food availability and safety in Singapore, the primary strategy of the country is to diversify food sources, support agricultural productivity through research and development, and to launch actions designed to reduce food waste.[20]

The Singaporean Food Authority (SFA) is fulfilling the food security tasks by enhancing "*the regulatory oversight over all food-related matters from farm to fork and further strengthen our food safety regime. This will also facilitate more business-friendly regulations. For example, by harmonising licensing standards and combining existing licenses for farmers, food manufacturers, food retailers, and food service operators, businesses will be able to adapt and transform their business models seamlessly and look at new ways to grow their operations.*"[21]

The sustainable production of 30% of Singapore's nutritional needs by 2030 is one of SFA's objective.[22] Due to limited space for conventional farming, the country relies on several innovations throughout the entire food chain.

Similar to the European Union, Singapore has authorisation and safety procedures for Novel Foods. The SFA considers novel food as food and food ingredients that do not have a history of safe use. The term "Novel food" also includes compounds that are chemically identical to naturally occurring substances but are produced through applications of technology, such as precision fermentation.[23] Similar to the EU provisions, food business operators dealing with precision fermentation need authorisation to sell Novel Foods on the Singaporean markets. These

[16] Tortajada and Hongzhou (2016), p. 1.

[17] Hamburg has a size of 755 qm^2 Hamburg (2019).

[18] Ludher (2016), p. 131.

[19] Ludher (2016), p. 133.

[20] Ludher (2016), p. 136.

[21] Agri-Food and Veterinary Authority of Singapore (2020).

[22] Singapore Food Agency (2022a).

[23] Requirements for the Safety Assessment of Novel Foods and Novel Food Ingredients Singapore Food Agency (2022b), p. 2.

companies must submit documents that prove the safety of their product for consumption.[24]

The Safety Assessment in Singapore for products derived through precision fermentation is more streamlined compared to other types of novel food. While a full safety assessment involving the submission of complete toxicity studies is not required, critical aspects of the safety assessment focus on product composition and characterisation, including major and minor components, as well as volatiles/non-volatiles.[25] The food business operator must demonstrate that the ingredients obtained through precision fermentation are chemically identical to its naturally occurring counterpart. The SFA provides a list of appropriate techniques/approaches that the food business operator can follow, such as Nuclear magnetic resonance spectroscopy, Infra-red spectroscopy, or Ultra-violet Spectroscopy.[26]

The SFA has already accepted products derived from precision fermentation as Novel foods, affirming their safety for consumption and allowancing their distribution.[27] Similar to the EU, misleading information is not allowed on the product labels. For precision fermentation products, this means that misrepresenting the method of ingredient production is prohibited. For example, milk proteins made using precision fermentation must not be labelled as milk proteins from cows.[28] Examples of precision fermented milk products available on the Singaporean market are ice cream and milk from Perfect Day.[29]

The legal landscape regulating food innovations in the **U.S.** differs significantly from the EU market, primarily due to varying approaches to technology and risk in food law. Unlike in the EU, the U.S. does not have a specific definition of "novel food", nor explicit legislation governing the market authorisation of these products or of "GM foods". Instead two pathways can be followed to place a new food (component) on the U.S. market: the Food Additive Safety Determination and the GRAS Determination,[30] with the latter being more advisable for precision fermented products.

Under the Federal Food, Drug, and Cosmetic Act (FD&C Act), food additives typically require pre-market approval by the Food and Drug Administration (FDA).[31] The term "food additive" encompasses *"any substance the intended use of which results or may reasonably be expected to result, directly or indirectly, in its becoming a component or otherwise affecting the characteristics of any food"*.[32] However,

[24] Singapore Food Agency (2023).

[25] Singapore Food Agency (2022b), p. 10.

[26] Full list see: Singapore Food Agency (2022b), p. 10 ff.

[27] Quan (2023).

[28] Quan (2023).

[29] Dang (2023).

[30] These procedures differ especially in relation to who accesses and reviews the scientific data and information provided for the determination Gaynor (2018).

[31] Center for Food Safety and Applied Nutrition (2005); Federal Food, Drug and Cosmetic Act (2022), Section 409; Shield et al. (2017).

[32] Federal Food, Drug and Cosmetic Act (2022), Section 201(s).

substances that are Generally Recognized as Safe (GRAS) by qualified experts for their intended use are exempt from this requirement. For substances used in food before 1958, GRAS status is granted based on common use experience. For all other substances, general recognition of safety must be established through scientific procedures and evaluated "based upon the application of generally available and accepted scientific data, information, or methods, which ordinarily are published, as well as the application of scientific principles".[33]

To determine whether a substance is GRAS for its intended use, technical evidence of its safety is required and must be accompanied by the confirmation that such evidence is generally known and accepted throughout the scientific community of those knowledgeable about the safety of the substance.[34] In practice, data showing the safety of the substance must be generated (according to the requirements listed in Title 21, Part 170, Subpart E of the Code of Federal Regulations (CFR)), and must successively be evaluated by a panel of scientific experts. The FDA describes the best practices for convening a GRAS panel in its guidance document as *"a panel of qualified experts who independently evaluate whether the available scientific data, information, and methods establish that a substance is safe under the conditions of its intended use in human food or animal food."*[35] The determination of GRAS status is the responsibility of the legal person placing the product on the market. In the first instance, GRAS determination is self-affirmed and independent of the FDA. The applicant provides data, convenes the panel of experts, and, upon agreement on the safety recognition, can use the substance in food and place it on the market without FDA's prior approval. However, the relevant documentation proving the GRAS status must be retained by the stakeholder to demonstrate that the procedure was lawfully conducted. FDA has the authority to challenge the GRAS status of substances used in foods after the products have reached the market.[36]

To avoid challenges, applicants can voluntarily notify FDA of their obtained GRAS claim by submitting a with substance description and characterisation, conditions of use, and information supporting the panel's recognition of safety. The FDA reviews the dossier within 180 (and, if necessary, additional 90) days and expresses an opinion on whether the reported data provide a sufficient basis for GRAS determination. If satisfied with the submitted documentation, FDA issues a so-called "no questions" letter to acknowledge agreement with the process and waives the possibility of further requests.[37]

The GRAS determination procedure offers several advantages for food innovation. While applicants must still generate data, which is generally published in peer-reviewed journals and made publicly available, the safety is assessed by panels of

---

[33] U.S. Food and Drug Administration (2022).

[34] Code of Federal Regulations. § 170.30 (2016); National Archives; Shield et al. (2017).

[35] U.S. Department of Health and Human Services (2022).

[36] Steele et al. (2016).

[37] Gaynor (2018); Steele et al. (2016).

experts. However, the process is less time-consuming, and the final decision is influenced by political factors less openly. The fact that products can enter the market after the independent GRAS determination provides a significant time advantage to companies and helps avoid saturating the capacities of the responsible authority by streamlining the procedures.

Already in September 2021 Clara Foods (now rebranded as "The EVERY Company") obtained a "no questions" letter from FDA regarding the GRAS status of their ClearEGG, a precision fermented non-animal egg white protein to be used as an ingredient for foods and beverages.[38]

In early 2020, the FDA issued a "no questions" letter as a response to the GRAS status of Perfect Day's precision fermented whey protein (β-lactoglobulin) under its intended conditions of use.[39] Although the use of this ingredient in infant formula and "products subject to regulation by the United States Department of Agriculture"[40] was explicitly excluded, it became the first PF-dairy alternative product considered safe for consumption in food and beverages on the US market. About two years later, in June 2022, Remilk followed suit by obtaining the self-affirmed GRAS status for their non-animal dairy protein.[41]

References

Agri-Food and Veterinary Authority of Singapore (2020) New Singapore Food Agency to Oversee Food Safety and Security. https://www.sfa.gov.sg/food-for-thought/article/detail/new-singapore-food-agency-to-oversee-food-safety-and-security. Accessed 16 June 2023

BRAIN Biotech AG (2023) Three examples of precision fermentation. https://www.brain-biotech.com/alternative-proteins/three-examples-of-precision-fermentation. Accessed 12 June 2023

Bundesministerium für Ernährung und Landwirtschaft (2018) Leitsätze für vegane und vegetarische Lebensmittel mit Ähnlichkeit zu Lebensmitteln tierischen Ursprungs: Neufassung vom 04. Dezember 2018 (BAnz AT 20.12.2018 B1, GMBl 2018 S. 1174). https://www.bmel.de/SharedDocs/Downloads/DE/_Ernaehrung/Lebensmittel-Kennzeichnung/LeitsaetzevegetrarischeveganeLebensmittel.pdf?__blob=publicationFile&v=5. Accessed 9 June 2023

Carlson S (2021) Re: GRAS Notice No. GRN 000967. https://www.fda.gov/media/152289/download. Accessed 16 June 2023

Center for Food Safety and Applied Nutrition (2005) CPG Sec 500.200 Food Additives –"GRAS": COMPLIANCE POLICY GUIDE (CPG). https://www.fda.gov/regulatory-information/search-fda-guidance-documents/cpg-sec-500200-food-additives-gras#:~:text=Section%20409%20of%20the%20Federal,substance%20may%20be%20safely%20used. Accessed 16 June 2023

Dang LH (2023) Precision fermentation adds value to food supply chain, not competition – Perfect Day. https://www.foodnavigator-asia.com/Article/2023/01/24/precision-fermentation-adds-value-to-food-supply-chain-not-competition-perfect-day#. Accessed 16 June 2023

[38] Carlson (2021); Watson (2021).

[39] Informa Markets (2020); Keller and Heckman (2019); Watson (2020).

[40] Keller and Heckman (2019).

[41] Ettinger (2022); Remilk Ltd (2022a, b).

Eder M (2023) Precision Fermentation simply explained. https://www.susupport.com/knowledge/fermentation/precision-fermentation#:~:text=Precision%20fermentation%20uses%20microbial%20hosts,pharmaceutical%20industry%20to%20produce%20biopharmaceuticals. Accessed 7 Aug 2023

Ettinger J (2022) Remilk's precision fermentation dairy is the second to earn U.S. GRAS Status. https://www.greenqueen.com.hk/remilk-u-s-gras-status/. Accessed 16 June 2023

European Parliament (2020) Amendments adopted by the European Parliament on 23 October 2020 on the proposal for a regulation of the European Parliament and of the Council amending Regulations (EU) No 1308/2013 establishing a common organisation of the markets in agricultural products, (EU) No 1151/2012 on quality schemes for agricultural products and foodstuffs, (EU) No 251/2014 on the definition, description, presentation, labelling and the protection of geographical indications of aromatised wine products, (EU) No 228/2013 laying down specific measures for agriculture in the outermost regions of the Union and (EU) No 229/2013 laying down specific measures for agriculture in favour of the smaller Aegean islands. https://eur-lex.europa.eu/legal-content/EN/TXT/?uri=CELEX:52020AP0289 Accessed 1 January 2024

Gaynor P (2018) How U.S. FDA's GRAS Notification Program Works. https://www.fda.gov/food/generally-recognized-safe-gras/how-us-fdas-gras-notification-program-works#authors. Accessed 16 June 2023

Goethem, B van (2016) Letter from the EC to EFSA, Subject: Fermentation products obtained from Genetically Modified Microorganisms (GMMs): Ref. Ares(2016)6016209 - 19/10/2016, SANTE/E3/TD/nn sante.ddg2.e.3(2016)6287877' (Brussels 16 October 2016)

Hamburg (2019) Hamburg in Zahlen: Fakten & Zahlen. https://www.hamburg.de/info/3277402/hamburg-in-zahlen/. Accessed 16 June 2023

Informa Markets (2020) Perfect Day's animal-free dairy protein earns Generally Recognized as Safe (GRAS) status. https://www.ingredientsnetwork.com/perfect-day-s-animalfree-dairy-protein-earns-news083312.html. Accessed 16 June 2023

Keller and Heckman (2019) GRAS Notice for Non-Animal Whey Protein from Fermentation by Trichoderma reesei. https://www.fda.gov/media/136754/download. Accessed 16 June 2023

Ludher EK (2016) Singapore's smart governance of food. In: Deakin M, Borrelli N, Diamantini D (eds) The governance of city food systems: case studies from around the world, pp 131–154

National Archives Code of Federal Regulations: A point in time eCFR system. https://www.ecfr.gov/current/title-21/chapter-I/subchapter-B/part-170/subpart-E?toc=1;. Accessed 16 June 2023

Pig Progress (2021) EU parliament decides fate of 'veggie meat'. https://www.pigprogress.net/home/eu-parliament-decides-fate-of-veggie-meat/#:~:text=In%20a%20recent%20vote%2C%20the,how%20products%20can%20be%20marketed. Accessed 16 June 2023

Quan TY (2023) Risk at a glance: new foods from fermentation. https://www.sfa.gov.sg/food-information/risk-at-a-glance/new-foods-from-fermentation

Remilk Ltd (2022a) Remilk Secures Self-Affirmed GRAS for Non-Animal, Real Dairy Protein. https://www.remilk.com/newsroom/remilk-secures-self-affirmed-gras-for-non-animal-real-dairy-protein-2. Accessed 16 June 2023

Remilk Ltd (2022b) Remilk Self-Affirms GRAS Status for Precision Fermentation Dairy. https://www.newprotein.net/news/remilk-self-affirms-gras-status-for-precision-fermentation-dairy. Accessed 16 June 2023

Schweizer E (2022) What consumers should ask about precision fermentation. https://www.forbes.com/sites/errolschweizer/2022/03/02/what-should-consumers-be-asking-about-precision-fermentation/. Accessed 7 Aug 2023

Shield J, Kuler J, Gurnani A (2017) Regulatory constraints on new product development and approval procedures in the United States. In: Developing new functional food and nutraceutical products. Elsevier, pp 123–148

Singapore Food Agency (2022a) Singapore Food Statistics 2022. https://www.sfa.gov.sg/docs/default-source/default-document-library/20230518-sfa-media-release%2D%2D-singapore-food-statistics-2022.pdf. Accessed 16 June 2023

Singapore Food Agency (2022b) Requirements for the safety assessment of novel foods and novel food ingredients. https://www.sfa.gov.sg/docs/default-source/food-import-and-export/Requirements-on-safety-assessment-of-novel-foods_26Sep.pdf. Accessed 16 June 2023

Singapore Food Agency (2023) Novel Food. https://www.sfa.gov.sg/food-information/novel-food. Accessed 16 June 2023

Sochirca N (2018) The European legal framework on vegan and vegetarian claims. EFFL 2018:514–521

Southey F (2022) Vegan 'steak' on the back burner? France suspends ban on 'meaty' terms for plant-based products. Accessed 16 June 2023

Steele EA, Breen C, Campbell E, Martin R (2016) Food regulations and enforcement in the USA. In: Reference module in food science. Elsevier

The EVERY Company (2023) EVERY. All Protein. No Animal. https://theeverycompany.com/. Accessed 12 June 2023

Tortajada C, Hongzhou Z (2016) Food policy in Singapore. In: Reference module in food science. Elsevier

U.S. Department of Health and Human Services (2022) Best practices for convening a GRAS panel: guidance for industry. https://www.fda.gov/media/109006/download. Accessed 16 June 2023

U.S. Food and Drug Administration (2022) Generally Recognized as Safe (GRAS). https://www.fda.gov/food/food-ingredients-packaging/generally-recognized-safe-gras. Accessed 16 June 2023

Watson E (2020) Perfect Day secures no objections letter from FDA for non-animal whey protein. https://www.foodnavigator-usa.com/Article/2020/04/14/Perfect-Day-secures-no-objections-letter-from-FDA-for-non-animal-whey-protein#:~:text=Perfect%20Day%2C%20a%20startup%20producing,non%2Danimal'%20whey%20protein. Accessed 16 June 2023

Watson E (2021) Clara Foods rebrands as The EVERY Company, launches 'nearly invisible' animal-free egg protein (without chickens). https://www.foodnavigator-usa.com/Article/2021/10/05/Clara-Foods-rebrands-as-The-EVERY-Company-launches-nearly-invisible-animal-free-egg-protein-without-chickens. Accessed 16 June 2023

Annex: Overview Interviews

F. Ronchetti et al., *The Regulatory Landscape in the EU for Dairy Products Derived from Precision Fermentation*, SpringerBriefs in Law,
https://doi.org/10.1007/978-3-031-49692-9

Interviewees	What is PF?	GMO	Novel food	Labelling	Bio	Vegan	Diary	Claims	Problems	Expectations
Interviewee I, Consultancy	Process where you control all parameters. Use of specific microorganisms	(+) Process itself, if over limit in end-product, otherwise →	(+) If no more GMO in end-Product	(+) Cheese, same components as "real" cheese	(−) because of GMO History	(+), because of animal welfare and sustainability	(+), It could be lactose free, but with other Milk proteins	(+) Product development, healthier cheese, (−) GI Future Dream	EU tough market, most consumers are sceptic	Better consumer information, choice between regular and PF-products
Interviewee II, Start-up	Bioidentical Proteins Produce Proteins in a Bioreactor during fermentation process. Later you filter these proteins, get a powder; purify it	(−) well purification; but depending on used proteins	(+) Casein	(−), Cheese because gene editing, no name yet because unsure about labelling	/	(+) but prefer to call product animal free for marketing reasons	(+) Want to be part of diary sector	Want higher nutritional value than regular cheese; (+), GI if bioreactor is based in the region	Other markets easier to access, e.g. in Singapore Cheese (+) FAQ: Labelling not many fermentation capacities	Labelling should be easier than for plant-based products. Commercialisation
Interviewee III, Consultancy	Difference to normal fermentation = vague; explanation not simple yet important; no microorganisms in the PF product "things are missing"	Thinks →	Thinks (+), not many applications yet	PF should be labelled descriptively since we are not in a position to get concrete; front pack labelling "unsexy", wishes memorandum for labelling	/	Animal Free fits best, vegan is unlikely	Unlikely claim	Creating healthier cheese = possible, but too early	Consumer acceptance Other markets more open, in US already PF products Political fights	Consumer safety important, early labelling agreements are mandatory for companies EFSA confirms safeness

Interviewees	What is PF?	GMO	Novel food	Labelling	Bio	Vegan	Diary	Claims	Problems	Expectations
Interviewee IV, Start-up	Important for raw material production, food itself manufactured via cell cultures. Targeted production of a specific molecule via microorganisms	Thinks (+), Because even CRISPR = GMO They do not want to access the market via food authorisation. Instead try via chemical approval	Thinks (–)	Working B2B, labelling not interesting, Genomic codes are too complex, only single proteins are produced, slightly different Genes than "real" cheese	/	V-Label says (+), even though animal cells are used; there will be always people who think it is not "vegan enough"; Unsure about animal free, maybe misleading	/	/	Authorisation too confusing, EU too complicated regulations.	/ (more problems than real expectations)
Interviewee V, Consultancy	PF is good for identity in sensory etc. But not for cellular composition. They have the same molecules, not same structure (cellular biology), Everyone is defining it differently. Dwayne: using single cells under specific conditions to produce a by-product. These cells have been genetically modified to produce sth they usually do not produce	GMO for cells to create needed proteins	Yes, unless you use Gene Editing	Not interested yet, not at the point to have those conversations, have to wait hoe people will call it, differs from legal name; Better: work with pictures on product to describe them	/	Better: Animal free	/	Possible: mixture from cell from different animals to create new product	PF must be more sustainable then the usual food production to be sustainable (P) Kosher/halal because not involved animal	Get access to European market

Interviewees	What is PF?	GMO	Novel food	Labelling	Bio	Vegan	Diary	Claims	Problems	Expectations
Interviewee VI, Consultancy	Technical approach: Single cell organisms to produce specific proteins. Fermentation + specific compounds = PF. Easier than engineering cells	Not sure, no idea. In US it is mostly no GMO, so he hopes the same applies for the EU.	Depending on the consultant you ask you get a different answer if NF or GMO, "break through" needed	Most consumers anyway don't know how their food is made, so why tell them how PF works? Maybe a disclaimer "new technology" enough? "Most Consumers don't care as long as it's delicious"	(−) Not conventional harvested	(+) no animals involved, Most important: Animal welfare. Also: No animal in the end product; but most companies do not advertise "vegan", feels unsexy, not delicious	Definition of dairy needed, instead: animal-free, slaughtered free	/	EU slow to understand importance of AltPro, Big Talent drain to US markets US and Singapore have clearer guidelines	EU has a lot of good research, hopefully gets faster soon
Interviewee VII, Consultancy		If not purification possible. See → If it (→) is in the end Product = GMO (produced with)	(+) as long as the end product is purified. If this is not working, we have a GMO. If the processing aid isn't in End Product = NF (produced from)	Maybe listed as ingredient the name of the product must be descriptive and not misleading. She things "Cheese with Casein derived from PF" is possible		(+) vegan in general, Non-Animal-Cheese, however there will be pushbacks,	(+) Companies want to be part of the diary sector, dairy proteins used	Keep it simple! First should start to be as close to conventional product as possible Maybe GI's if process in the region, very far future	More consumer education for PF Wish for more clarification/ information from EFSA	NF approval in EU, Revolution in the way we eat

SpringerBriefs in Law and Regulation & Innovation present summaries of cutting-edge research and practical applications across the wide spectrum of regulatory law. Regulation has established itself as an important component of the legal system. It aims to achieve the common good in the interaction between and among administration and private individuals and occurs alongside the other central (administrative) legal functions of limiting state power to protect freedom, averting dangers, regulating the direct provision of state services and the general provision of legal rules for social development. SpringerBriefs in Law and Regulation & Innovation are dedicated to the understanding, systematic recording and dogmatic and empirical processing of such regulatory regimes. It is open for basic questions, central elements and new developments in the regulatory law. The focus is on legal regimes to secure results that promote the common good from and through economic competition, such as those found in network economies, data law, and increasingly in the food as well as in the health sector etc. Featuring compact volumes of 50 to 125 pages, the series covers a range of content from mainly professionals but also academics.

GPSR Compliance
The European Union's (EU) General Product Safety Regulation (GPSR) is a set of rules that requires consumer products to be safe and our obligations to ensure this.

If you have any concerns about our products, you can contact us on

ProductSafety@springernature.com

In case Publisher is established outside the EU, the EU authorized representative is:

Springer Nature Customer Service Center GmbH
Europaplatz 3
69115 Heidelberg, Germany

www.ingramcontent.com/pod-product-compliance
Ingram Content Group UK Ltd.
Pitfield, Milton Keynes, MK11 3LW, UK
UKHW020704041026
15590UKWH00001B/456
* 9 7 8 3 0 3 1 4 9 6 9 1 2 *